ON THE APPLICATION OF DATA ASSIMILATION IN REGIONAL COASTAL MODELS

T0138775

On the Application of Data Assimilation in Regional Coastal Models

DISSERTATION
Submitted in fulfilment of the requirements of
the Board for Doctorates of Delft University of Technology
and of the Academic Board of the International Institute for Infrastructural,
Hydraulic and Environmental Engineering for the Degree of DOCTOR
to be defended in public
on Monday, 8 February 1999 at 10:30 h

by
RAFAEL CAÑIZARES TORRE

born in Madrid, Spain
Master of Engineering in Civil Engineering, Polytechnical University of Madrid
Master of Science in Hydraulic Engineering, International Institute for
Infrastructural Hydraulic and Environmental Engineering, Delft

A.A. BALKEMA / ROTTERDAM / BROOKFIELD / 1999

To Eugenia

ABSTRACT

Data assimilation is a methodology which can optimise the extraction of reliable information from observations and combine it with, or assimilate it in, numerical models. The development of both the observation techniques and the numerical models (associated with the increase in computer speed and memory) has made of data assimilation a very important and attractive field within oceanography, meteorology and engineering.

This work deals with the integration and further development of existing computationally efficient data assimilation techniques for large scale regional coastal models. Two suboptimal schemes of the Kalman filter are described. The first scheme is a reduced rank square root (RRSQRT) approximation of the extended Kalman filter, which approximates the error covariance matrix by one of a lower rank. The second is the ensemble Kalman filter (EnKF), which calculates the error statistics using a Monte Carlo method. Both techniques can approximate the results obtained from the Kalman filter but at a much lower cost.

The performance of these methods is compared in a twin test using a two-dimensional shallow water equation model. The performance of the RRSQRT filter for forecasting and hindcasting has been tested in the standard two-dimensional model and in a special version that computes the shallow water equations in areas of different resolution (nesting). The final part of this work presents the initial tests of the implementation of the RRSQRT in a three-dimensional model.

Contents

1

Chapter 1

Introduction

For many centuries, the process of understanding natural phenomena was basically restricted to just two sub-processes: observing and reasoning. Based on the results of this process, it was possible to define physical laws, to describe those phenomena, such as those which later came to be expressed by differential equations. These equations could not usually be solved directly. Instead, discrete numerical methods were introduced to provide an 'almost' perfect solution if provided with the correct forcing and parameters.

With the creation of computers, a large part of the world's research effort was directed to the development of numerical models. Numerical models are able to solve systems of equations of unimaginable size and complexity. The qualitative physical knowledge concentrated in a set of physical laws can be transformed into quantitative knowledge of physical phenomena. However, in spite of the rapid improvement of numerical models, the modelled phenomena are sometimes very different from the observed.

It is well known that numerical models are not perfect. Due to the discretisation used in a numerical model, sub-grid scale effects exist which cannot be considered. Moreover, although the deterministic part of the phenomenon can be quite accurately modelled, this is not the case for the stochastic part. These two simplifications, together with the error in the model parameterisation (because most of the model parameters cannot be measured) are the main sources of errors inherent in a numerical model.

On the other hand, in order to obtain a solution, the model has to be provided with the necessary external forces, for example the open boundary conditions and the meteorological forcing. The uncertainty in these forces is usually more significant than the errors in the numerical model. In order to model certain phenomena the forcing that causes it has to be specified. When the error in the forcing is significant the model simulates a different phenomena, namely one that corresponds to the specified (and false) forcing.

The application of regional coastal models to storm surge modelling, current forecasting and flood risk forecasting requires highly accurate model results. The application has to be validated in the sense of establishing the ability of the system and its components to deal with the physical phenomena under study. After setting-up the model, it has to be calibrated by tuning the system parameters in order to narrow the gap between the results of the simulation and the observations. If the differences between the simulated and the observed phenomena are caused by model simplifications and/or errors in the forcing (for example, open boundary conditions, meteorological forcing) no significant improvements can be achieved during the calibration.

In order to account for these errors directly in the numerical model, the model has to be integrated with a data assimilation method.

1.1   Data assimilation

Data assimilation is a methodology which can optimise the extraction of reliable information from observations and combine it with, or assimilate it into, numerical models. Data assimilation has been recently included within oceanography in much the same way as it is already used in the related fields of meteorology and engineering.

Data assimilation provides an estimation of the state of the system by integration of observed data with a dynamical model of the system. The use of the dynamical model has a major importance for forecasting as well as for propagating the system state forward in time.

The concept of data assimilation is directly associated with the concept of errors. The three components of a data assimilation system; -the measurements, the dynamical model and the data assimilation technique- each have associated errors. The set of observations contains errors, such as instrument noise and due to the interpretation of measured variables as state variables, etc. The dynamical models are imperfect, containing errors due to unresolved physics and sub-grid scale errors. Approximations and simplifications in the data assimilation methods also produce errors in linking the model and the observations. A well constructed data assimilation method has to agree with all the observations within data errors and has to satisfy the dynamical model within model error limits. An indication of the importance of the relationship between model and measurement errors can be gauged already from this characterisation.

1.2   Review of different data assimilation methods

The use of data assimilation in coastal hydrodynamical modelling is quite recent. The techniques applied have been directly used or adapted from other fields such as engineering, meteorology and oceanography. An excellent review of data assimilation techniques applied in Oceanography can be found in the paper of Ghil and Malanotte-Rizzoli [28]. In general, data assimilation techniques can be divided into two main groups the roots of which are originally to be traced to estimation theory and control theory. Hereafter, an attempt is made to review the most-well known and applied data assimilation techniques.

1.2.1 Techniques from estimation theory

Estimation theory is usually associated with a sequential assimilation process where the expected error of the estimate is minimised in terms of model and data errors statistics. In sequential assimilation the assimilating model is integrated over the time interval and during this period observations are distributed. The model produces a predicted system state and, whenever observations are available, this is updated or corrected using those new observations. The updated model using the observations is known as the analysis. The new model integration uses the updated state as the new initial condition. In sequential assimilation, observations only have influence on the estimated state at later times, not at previous times. The most widely used sequential data assimilation techniques applied in oceanography as well as in meteorology are outlined here.

Direct insertion is probably the simplest technique. Whenever observations of the state are available, the counterpart variables in the system state are substituted by the observations. Haines [29] applied this technique in a quasi-geostrophic model. He directly inserted potential vorticity fields into the top layer, so producing significant corrections in the deeper layers between the assimilation times.

Blending technique is a simplified and localised version of optimal interpolation with purely empirical weights. Where observations are available, the forecasted variable is replaced by a new variable obtained as a result of blending the forecasted and the observed variable. When the forecasted variable is directly substituted by the observed variable the method is called direct insertion. -Thus, in the combination of both variables, zero weight is given to the forecasted and one to the observed.-

Optimal interpolation is the data assimilation technique most commonly used for numerical weather prediction (e.g. Lorenc [43], Daley [14]). It is very useful when dealing with large sets of heterogeneous observations. Optimal Interpolation is only based on statistical assumptions and not on the model dynamics, but still has a better definition of the error distribution than direct insertion. The method (described in detailed in Gelb [27] or Daley [14]) initially assumes that the error covariances of the model forecast and the observations are known. It provides computable expressions of the analysis error covariance matrix that allows the calculation of error bars. The error covariance is not propagated forward in time; instead it generally has a fixed analytical spatial structure (DeMey [16]). Some examples in oceanography can be found in Rienecker and Miller [55] and De Mey [15], [16]. Another more recent application of optimal interpolation in the field of air pollution, Zhang [68], applies two different Kriging techniques in order to describe the spatial distribution of the errors.

Nudging or Newtonian relaxation is based on Anthes [3] who used the method to assimilate data into an atmospheric model. It consists of a dynamical model relaxation towards the observations. To achieve that, a term proportional to the difference between observed and calculated variables is added to the state continuous evolution equation. Haines *et al.* [30] show the results of the nudging method in a shallow water model for a twin experiment. Another application of nudging in a different field was carried out by Sokolov *et al.* [56]. They assimilated temperature and salinity data into the transport equations of a 3D baroclinic hydrodynamic model.

The Kalman filter is one of the most explored data assimilation techniques, it has been applied in many geophysical fields. Originally developed by Kalman in 1960 [37] it is an efficient data assimilation method that explicitly accounts for the dynamic propagation of errors in the model. In the case of linear models and under the assumption of known statistics of the system and the measurement errors, the Kalman filter provides an optimal state of the system in terms of minimum estimation error variance. In the case of weakly non-linear model dynamics, an approximate Kalman filter algorithm (extended Kalman filter) in which the error propagation is based on a statistical linearisation of the model equations, can be adopted (Jazwinsy [35]). Despite the numerous applications of the Kalman filter that can be found in the literature, applications to large systems are scarcely feasible due to the huge computational burden involved. In recent years, several so-called suboptimal schemes (SOS's) (Todling and Cohn [59]) have been formulated. These schemes make use of different approximations of the error covariance modelling in order to reduce the computational burden of a Kalman filter application.

Most sub-optimal schemes are based on either a simplification of the model dynamics used in propagating the errors or an approximation of the error covariance matrix. Within the first group, Dee [17] applied a simplified Kalman filter in an atmospheric model where mass error propagation was obtained by simple advection and the wind error propagation was subsequently evaluated by imposing a geostrophic balance in the momentum equation. Another approach presented by Fukimori and Malanotte-Rizzoli [26] consists of propagating the error covariance matrix in a coarser model than the one used for the propagation of the original model. Cohn and Todling [13] proposed a method where a reduced rank approximation of the singular value decomposition of the tangent linear operator is applied in the extended Kalman filter.

Regarding the second group, the one using an approximation of the error covariance matrix, researchers have followed different directions. Heemink [31] developed a steady (time invariant) Kalman filter where the Kalman filter can be calculated off-line yielding a significant reduction of the computational burden. This technique has been successfully applied for storm surge forecasting problems in Heemink [33], Vested *et al.* [65], Bolding [5] and Heemink *et al.* [34]. Some of the important restrictions of this technique are a time invariant error statistics and fixed measurement positions. In order to use a time invariant Kalman filter, an efficient scheme based on a reduced rank approximation of the error covariance matrix has been introduced. The error covariance matrix is approximated by a matrix of lower rank, which contains only the most significant eigenvectors of the full matrix. Verlaan and Heemink [63] and [64] used this approach together with a square root factorisation while Cohn and Todling [13] used a Lanczos algorithm for the eigenvalue decomposition.

Monte Carlo simulations. In the case of strongly non-linear dynamics the Extended Kalman filter may fail. Evensen [21] found that the extended Kalman filter for a non-linear quasi-geostrophic ocean model resulted in an unbounded error covariance growth. The Extended Kalman filter can be improved by including higher order moments in the error covariance approximation, increasing the computational requirements significantly. Evensen [22] proposed an approach, the ensemble Kalman filter, which is based in a Monte Carlo simulation and can resolve non-linearities in the error propagation. In this method the error covariance is represented by an ensemble of possible states that are propagated according to the fully non-linear dynamics of the system. This method has been applied in a quasi-geostrophic ocean model for assimilation of altimeter data by Evensen and van Leeuwen [25].

1.2.2 Techniques from control theory

These methods are based on the variational principle and they can be considered more as a smoothing process. Variational assimilation adjusts the model solution over the entire assimilation period to all the observations available at the same period. Estimation at all times are influenced by all the observations during the assimilation period. The most widely applied variational method is the adjoint method (Talagrand and Coutier [57], Thacker and Long [58]) which can be considered as a very efficient method for direct minimisation with strong constraints. The method computes the gradient of a quadratic function with respect to the variables to be adjusted, for the exact solution of the model equations. The exact model solution is obtained by propagating the difference between the trajectory of the model and the observations, backward in time with the adjoint equations. In the area of coastal modelling, the adjoint method has been applied for off-line estimation of model parameters (Lardner *et al.* [39], Verlaan [60]). The possibility of using a weak-constraint of the cost function in order to account for model errors has been recently applied by Eknes and Evensen [20].

Some important drawbacks of the variational methods are the difficulty of applying such methods in on-line forecasting procedures and the complexity of the adjoint models, which makes the development of the software a very complicated task.

1.3   Thesis outline

This thesis focuses on the integration and development of existing computationally efficient data assimilation techniques for large scale regional coastal models. The thesis makes a contribution to the future development of operational models for the Danish waters.

Chapter 2 describes the deterministic hydrodynamic models used during the thesis. Both models, the two-dimensional and the three-dimensional shallow water equation models, are introduced.

In Chapter 3, the Kalman filter is presented and described for the case of linear as well as non-linear models. Two approximations of the Kalman filter, the reduced rank square root filter (RRSQRT) and the ensemble Kalman filter (EnKF), are described. The specific implementation of these techniques in the two-dimensional shallow water equation model is also investigated during this chapter.

In Chapter 4, the inter-comparison and the sensitivity study of the RRSQRT and the EnKF is presented. The results from different twin tests dealing with errors in the open boundary conditions and in the meteorological forcing terms are presented.

Chapter 5 presents a real-world implementation of the RRSQRT filter in the two-dimensional hydrodynamic model. One of the most studied region of the world, the North Sea, has been selected for this application. The filter capabilities for hindcasting and forecasting have been studied and the results are presented.

In chapter 6, the application of the filter in a hydrodynamic model that computes simultaneously areas of different resolution is presented. The implementation has been tested in a regional model of the North Sea and the Baltic Sea. Two cases, the case with a unique area and the one obtained adding to the former a fine resolution area in the Danish inner waters, are compared.

In Chapter 7 the implementation of the RRSQRT filter on the barotropic part of a three-dimensional hydrostatic model is presented. The implementation is tested in two specific cases within the same twin test used in chapter 2, using the systematic error at the open boundary and the systematic error in the meteorological forcing.

Conclusions are finally given in Chapter 8.

Chapter 2

The Deterministic Hydrodynamic model

2.1   Introduction

The most widely applied shallow water equations models are still two-dimensional depth averaged. Due to this fact, the main part of this study deals with the application of data assimilation methods in two-dimensional models. Initially, a sequential data assimilation technique has been applied to the hydrodynamic module of the MIKE 21 (Abbott *et al.* [1]) modelling system. The purely deterministic version of this model has been already used for forecasting, so that the combination with data assimilation techniques was a necessary step forward to improve the forecast.

Nowadays, three-dimensional shallow water equation models are becoming more commonly used as a tool for coastal area modelling. Therefore, at the last stage of this study the application of data assimilation in a three-dimensional hydrostatic model has been carried out and tested.

This chapter describes the model equations and the numerical solution applied for both models the two- and the three-dimensional models.

2.2   The 2D shallow water equations model

The MIKE 21 is a general numerical modelling system that has been largely applied for simulation of water levels and flows in estuaries, bays and coastal areas. Among hundreds of different applications, this model has been applied to an operational system for prediction and forecasting of water levels and currents in the Danish Belts. The purely deterministic version was used for this purpose, so that its combination with data assimilation techniques was the next natural attempt for improving the forecast. The MIKE 21 modelling system solves the vertically-integrated equations

11

of continuity and conservation of momentum in two horizontal directions (see Vested *et al.* [65]).

The equations that are solved are those of:

continuity:

$$\frac{\partial \zeta}{\partial t} + \frac{\partial p}{\partial x} + \frac{\partial q}{\partial y} = S - e \tag{2.1}$$

x-momentum:

$$\frac{\partial p}{\partial t} + \frac{\partial}{\partial x}\left(\frac{p^2}{h}\right) + \frac{\partial}{\partial y}\left(\frac{pq}{h}\right) + gh\frac{\partial \zeta}{\partial x} + \frac{g\frac{p}{h}\sqrt{\frac{p^2}{h^2} + \frac{q^2}{h^2}}}{C^2}$$
$$-fVV_x - \frac{h}{\rho_w}\frac{\partial}{\partial x}(P_a) - \Omega q - E\left(\frac{\partial^2 p}{\partial x^2} + \frac{\partial^2 p}{\partial y^2}\right) = S_{ix} \tag{2.2}$$

and y-momentum:

$$\frac{\partial q}{\partial t} + \frac{\partial}{\partial y}\left(\frac{q^2}{h}\right) + \frac{\partial}{\partial x}\left(\frac{pq}{h}\right) + gh\frac{\partial \zeta}{\partial y} + \frac{g\frac{q}{h}\sqrt{\frac{p^2}{h^2} + \frac{q^2}{h^2}}}{C^2}$$
$$-fVV_y - \frac{h}{\rho_w}\frac{\partial}{\partial y}(P_a) - \Omega p - E\left(\frac{\partial^2 q}{\partial x^2} + \frac{\partial^2 q}{\partial y^2}\right) = S_{iy} \tag{2.3}$$

where:

$\zeta(x,y,t)$	water surface level above datum (m)
$h(x,y,t)$	water depth (m)
$p(x,y,t)$	flux density in the x-direction $(m^3 s^{-1} m^{-1})$
$q(x,y,t)$	flux density in the y-direction $(m^3 s^{-1} m^{-1})$
S	source magnitude per unit horizontal area $(m^3 s^{-1} m^{-2})$
S_{ix}, S_{iy}	source impulse in x and y directions$(m^3 s^{-1} m^{-2}.ms^{-1})$
e	evaporation rate (ms^{-1})
g	gravity (ms^{-2})
C	Chézy resistance No. $(m^{1/2} s^{-1})$
f	wind friction factor
$V, V_x, V_y(x,y,t)$	wind speed and components in x and y directions (ms^{-1})
$P_a(x,y,t)$	barometric pressure $(kgm^{-1}s^{-2})$
ρ_w	density of water (kgm^{-3})
Ω	Coriolis coefficient (latitude dependent) (s^{-1})
$E(x,y)$	eddy "viscosity" coefficient $(m^2 s^{-1})$
x,y	space co-ordinates (m)
t	time (s)

Correction terms can be introduced on the flux densities in order to account for markedly non-rectangular velocity distributions, but these were not used in the present work. At closed boundaries, the flow perpendicular to the boundary is set to zero. At open boundaries, the surface elevation along the boundary and/or the flux through the boundary have to be prescribed. Using these boundary conditions and defining initial values for surface elevations and flux densities, the system of equations 2.1 - 2.3 forms a well-posed boundary value problem.

2.2.1 Numerical Solution

The equations 2.1 - 2.3 are solved by implicit finite difference techniques. These techniques are a fractioned-step technique combined with an Alternating Direction Implicit (ADI) algorithm to avoid the necessity of iteration. Second-order accuracy is ensured through the centering in time and space of all derivatives and coefficients.

The variables are defined on a space-staggered rectangular grid with elevation and fluxes midway between grid points (Leendertse [40]). The three equations are time centered at $t + \frac{1}{2}$ and they are solved in one dimensional sweeps alternating between the x and y directions:

1. X-sweep. The continuity and x-momentum equations are solved, taking ζ from t to $t + \frac{1}{2}$ and the flux density p from t to $t+1$. Terms involving the flux density in the y-direction q use the already known values at levels $t - \frac{1}{2}$ and $t + \frac{1}{2}$. The mass and momentum equations for the x-sweep for a sequence of grid points are:

$$AMA_j p_{j-1}^{n+1} + BMA_j \zeta_j^{n+\frac{1}{2}} + CMA_j p_j^{n+1} = DMA_j \qquad |_k$$
$$AMO_j \zeta_j^{n+\frac{1}{2}} + BMO_j p_j^{n+1} + CMO_j \zeta_{j+1}^{n+\frac{1}{2}} = DMO_j \qquad |_k \qquad (2.4)$$

2. Y-sweep. The continuity and y-momentum equations are solved, taking ζ from $t + \frac{1}{2}$ to $t + 1$ and the flux density q from $t + \frac{1}{2}$ to $t + \frac{3}{2}$. Terms involving the flux density in the x-direction p use the values just calculated during the x-sweep at t and $t + 1$. The mass and momentum equations for the y-sweep for a sequence of grid points are

$$AMA_k q_{k-1}^{n+\frac{3}{2}} + BMA_k \zeta_k^{n+1} + CMA_k q_k^{n+\frac{3}{2}} = DMA_k \qquad |_j$$
$$AMO_k \zeta_k^{n+1} + BMO_k q_k^{n+\frac{3}{2}} + CMO_k \zeta_{k+1}^{n+1} = DMO_k \qquad |_j \qquad (2.5)$$

The coefficients $AMA, AMO, BMA, BMO, CMA, CMO, DMA$ and DMO in the previous equations, are all expressed in known quantities. The application of the implicit finite difference scheme results in a tridiagonal system of equations for each grid line in the model. The solution is obtained by inverting the tridiagonal matrix using the Double Sweep (DS) algorithm (see for example Abbot and Minns [2]), which corresponds to a very fast and accurate form of Gauss elimination.

2.3   Initial definition of the state vector in MIKE 21

The state variables used in MIKE 21 are: surface elevations ζ, and flux densities p and q in the x and y directions respectively, defined on a staggered C-grid in the $x - y$ space. The solution technique in MIKE 21 involves the use of variables at three time steps, while the Kalman filter is based on a recursive two time step formulation. However, to represent this scheme in a two time steps formulation in order to be used in the Kalman filter, the flux density q at time steps $k + \frac{1}{2}$ and $k - \frac{1}{2}$ are included in the state vector. The state vector can then be written as

$$
x_k = \begin{pmatrix} \zeta_k \\ p_k \\ q_{k+\frac{1}{2}} \\ q_{k-\frac{1}{2}} \end{pmatrix} \tag{2.6}
$$

Further specifications about the state vector definition are presented in chapter 3.

2.4   The 3D shallow water equations model

The MIKE 3 HS is a general numerical modelling that solves the governing three-dimensional equations under the assumption of hydrostatic pressure. The consequence of this assumption is that the momentum equation for the vertical is not solved obtaining the vertical velocity directly from the continuity equation.

The vertical grid is divided in a number of layers and it is defined using a general vertical co-ordinate (see Buchard [6]). For each layer, it is possible to define the layer-integrated equations from the primitive equations. After a transformation to a vertical general grid, the layer-integrated equations can be written (see Pietrazk *et al.* [51]):

continuity:

$$
\frac{\partial \Delta h}{\partial t} + \frac{\partial \Delta p}{\partial x} + \frac{\partial \Delta q}{\partial y} + \bar{w}_{top} - \bar{w}_{bot} = 0 \tag{2.7}
$$

x-momentum:

$$\frac{\partial \Delta p}{\partial t} + \frac{\partial}{\partial x}\left(\frac{\Delta p \Delta p}{\Delta h}\right) + \frac{\partial}{\partial y}\left(\frac{\Delta p \Delta q}{\Delta h}\right) +$$

$$(\bar{w}u)_{top} - (\bar{w}u)_{bot} - \tau_{top}^x + \tau_{bot}^x - \Delta h F_x + \Omega \Delta q = -\frac{1}{\rho_o}\frac{\partial}{\partial x}(P_{pres}) \quad (2.8)$$

y-momentum:

$$\frac{\partial \Delta q}{\partial t} + \frac{\partial}{\partial y}\left(\frac{\Delta q \Delta q}{\Delta h}\right) + \frac{\partial}{\partial x}\left(\frac{\Delta p \Delta q}{\Delta h}\right) +$$

$$(\bar{w}v)_{top} - (\bar{w}v)_{bot} - \tau_{top}^y + \tau_{bot}^y - \Delta h F_y + \Omega \Delta p = -\frac{1}{\rho_o}\frac{\partial}{\partial y}(P_{pres}) \quad (2.9)$$

and the layer-integrated equations from salinity and temperature:

$$\frac{\partial}{\partial t}(\Delta h \Delta S) + \frac{\partial}{\partial x}(\Delta p \Delta S) + \frac{\partial}{\partial y}(\Delta q \Delta S) + (\bar{w}\Delta S)_{top} - (\bar{w}\Delta S)_{bot} = 0 \quad (2.10)$$

$$\frac{\partial}{\partial t}(\Delta h \Delta T) + \frac{\partial}{\partial x}(\Delta p \Delta T) + \frac{\partial}{\partial y}(\Delta q \Delta T) + (\bar{w}\Delta T)_{top} - (\bar{w}\Delta T)_{bot} = 0 \quad (2.11)$$

where:

$\zeta(x,y,t)$	water surface level above datum (m)
$\Delta p(x,y,t)$	Layer-integrated flux density in the x-direction $(m^3 s^{-1} m^{-1})$
$\Delta q(x,y,t)$	Layer-integrated flux density in the y-direction $(m^3 s^{-1} m^{-1})$
$\Delta h(x,y,t)$	Layer thickness (m)
$\bar{w}$	vertical velocity perpendicular to moving layer interface (ms^{-1})
u	velocity at top of the layer interface in the x-direction (ms^{-1})
v	velocity at top of the layer interface in the y-direction (ms^{-1})
$\Delta S(x,y,t)$	Layer-integrated salinity
$\Delta T(x,y,t)$	Layer-integrated temperature
τ	Layer interface shear stress
F	friction term
$P_{pres}(x,y,t)$	pressure $(kg m^{-1} s^{-2})$
ρ_o	reference density $(kg m^{-3})$
Ω	Coriolis coefficient (latitude dependent) (s^{-1})
x,y	space co-ordinates (m)
t	time (s)

The horizontal shear stress term, the shear stress term and the pressure gradient for the x-direction can be written as:

$$F_x = \frac{\partial}{\partial x}\left[2\nu_{T,H}\frac{\partial u}{\partial x}\right] + \frac{\partial}{\partial y}\left[\nu_{T,H}\left(\frac{\partial u}{\partial y} + \frac{\partial v}{\partial x}\right)\right] \tag{2.12}$$

$$\tau_{top}^x = \nu_{T,V}\left(\frac{\partial u}{\partial x}\right)_{top} \tag{2.13}$$

$$-\frac{1}{\rho_o}\frac{\partial}{\partial x}(P_{pres}) = -g\Delta h\frac{\rho(\zeta)}{\rho_o}\frac{\partial \zeta}{\partial x} - \frac{g}{\rho_o}\int_{bot}^{top}\frac{\partial \rho}{\partial x}\partial z \tag{2.14}$$

The horizontal shear stress and the pressure gradient terms refers to the Cartesian co-ordinate system since it is not possible to make an exact transformation due to the non-linearities in the terms.

2.4.1   Numerical solution

This model uses a very similar technique (an alternating direction implicit ADI) to the one used in the two-dimensional model for solving the continuity and momentum equations. The only difference is the inclusion of the vertical velocity calculation after the x-sweep and the y-sweep. In comparison with MIKE 21, the horizontal convective and friction terms are now explicit, resulting in no extra terms in the horizontal direction. The vertical convective and friction terms are now implicit. The solution procedure (Pietrazk et al. [51]) essentially consists of five steps in either of the sweep directions. The x-sweep is explained hereafter, therefore all the equations refer to a given location in the y-direction.

The first step consists of the calculation of the coefficients of the layer-integrated momentum equation along the specified k-line

$$AL_{j,l}\zeta_j^{n+\frac{1}{2}} + DOL_{j,l}\Delta p_{j,l-1}^{n+1} +$$
$$BL_{j,l}\Delta p_{j,l}^{n+1} + UPL_{j,l}\Delta p_{j,l+1}^{n+1} + CL_{j,l}\zeta_{j+1}^{n+\frac{1}{2}} = DL_{j,l} \tag{2.15}$$

this is followed by a reduction of the equation through a partial elimination of the layer integrated momentum equation over a column of water. This results in the following reduced tridiagonal equation with new values for the coefficients

$$AL_{j,l}\zeta_j^{n+\frac{1}{2}} + \Delta p_{j,l}^{n+1} + CL_{j,l}\zeta_{j+1}^{n+\frac{1}{2}} = DL_{j,l} \tag{2.16}$$

The second step consists of a summation of the reduced equation over depth,

$$\sum_{l=0}^{L} AL_{j,l}\zeta_j^{n+\frac{1}{2}} + \sum_{l=0}^{L} \Delta p_{j,l}^{n+1} + \sum_{l=0}^{L} CL_{j,l}\zeta_{j+1}^{n+\frac{1}{2}} = \sum_{l=0}^{L} DL_{j,l} \qquad (2.17)$$

Since the surface elevation ζ is independent of l, this can be re-written as

$$A_j\zeta_j^{n+\frac{1}{2}} + p_j^{n+1} + C_j\zeta_{j+1}^{n+\frac{1}{2}} = D_j \qquad (2.18)$$

where:

$$p_j^{n+1} = \sum_{l=0}^{L} \Delta p_{j,l}^{n+1} \quad ; \quad A_j = \sum_{l=0}^{L} AL_{j,l}$$

$$C_j = \sum_{l=0}^{L} CL_{j,l} \quad ; \quad D_j = \sum_{l=0}^{L} DL_{j,l}$$

which is identical to the MIKE 21 formulation for the depth integrated momentum equation. The third step involves the calculation of the coefficients for the depth-integrated continuity equation. The fourth step uses the same algorithm as used in the MIKE 21 case to obtain the depth integrated solution, from the above equation and coefficients. Having the depth-integrated solution, and in particular the surface elevation the flux can be calculated for each layer. Rearranging equation 2.16 yields the following explicit formulation

$$\Delta p_{j,l}^{n+1} = DL_{j,l} - AL_{j,l}\zeta_j^{n+\frac{1}{2}} - CL_{j,l}\zeta_{j+1}^{n+\frac{1}{2}} \qquad (2.19)$$

Finally in the fifth step, the layer integrated continuity equation is used to obtain the vertical velocities. Simply starting from the bottom with the boundary condition $w=0$ and solving sequentially for each layer.

The advection of temperature and salinity in the model are calculated using the ULTIMATE scheme from Leonard [41] (see also Pietrazk [50]).

For further details about this model the reader is referred to Pietrazk *et al.* [51].

Chapter 3

Kalman filter techniques

Probably, the most well known and widely used sequential data assimilation technique is the Kalman filter. It was introduced by Kalman [37] in 1960 within the field of control theory. The Kalman filter is the statistically optimal method for the sequential assimilation of data into linear numerical models and provides an estimate of the state of the system at the current time step based on all measurements of the system available up to and including the current time. Typical applications of the Kalman filter in linear models for storm surge forecast can be found in Heemink [31], Vested *et al.* [65] and Pietrazk and Bolding [49]. In these cases, a linearised hydrodynamic model was coupled to a stationary Kalman filter which could be calculated off-line. The Kalman filter can also be applied to weakly non-linear systems using the so-called extended Kalman filter (Kalman and Bucy [38]). The computational burden associated with the propagation of the error statistics forward in time, however makes the application of the extended Kalman filter to large systems untenable. As a result, new Kalman-filter-based algorithms have been developed. Within these new algorithms, a first approach posed the problem of propagating the error statistics in a more efficient manner. This leads to the so called Suboptimal Schemes (SOS) which can in turn be grouped into two approaches: simplification of model dynamics (Todling and Cohn [59]) which makes use of a state-transition matrix with lower rank; and simplification of covariance modelling as the Reduced Rank Square Root Filter (Verlaan and Heemink [63]) which approximates the error covariance matrix by a matrix of lower rank. A different approach for solving this problem is the ensemble Kalman filter of Evensen [23] where the error statistics are calculated using Monte Carlo methods.

3.1   Linear filtering

Many descriptions and derivations of the Kalman filter can be found in the literature (see for example Ghil and Malannotte-Rizzoli [28] and Evensen [23]).

A state vector representation of a stochastic discretized linear model can be written as

$$x_{k+1} = A_k x_k + B_k u_k + \Lambda_k w_k \tag{3.1}$$

where $x_k \epsilon \Re^n$ is the state vector at time t_k with n the number of model variables (number of unknowns). Matrices A_k and B_k represent the model dynamics. The model forcing is represented by a matrix u_k. $w_k \epsilon \Re^m$ is the unbiased model noise, which represents the errors in the numerical model, the mathematical model and such model-forcing features as the boundary conditions and the wind forcing. The number of points at which the system noise is defined is represented by m. For example, if some noise is introduced at the boundaries and in the momentum equation, m represents the number of noise points at the boundaries plus the number of noise points for the momentum equation (i.e. for the x and y directions). $\Lambda_k \epsilon \Re^{n \times m}$ is the interpolation matrix between the noise grid and the model grid. If $n = m$, i.e. the noise is defined on the model grid, the interpolation matrix becomes unitary. Equation (3.1) represents the propagation of the system one step forward in time.

When multiple measurements are available, the measurement vector $z_k \epsilon \Re^p$ is related to the state vector x_k through the measurement equation

$$z_k = H_k x_k + v_k \tag{3.2}$$

where $H_k \epsilon \Re^{p \times n}$ contains the relations between measurements and model variables, p is the number of measurements and $v_k \epsilon \Re^p$ contains the unbiased measurement noise.

The Kalman filter provides a linear unbiased estimate of the state vector that combines model forecast and measurements while the inherent uncertainties are taken into account. In order to obtain an optimal solution in terms of a minimum error covariance, the following assumptions have to be made (Ghil and Malannotte-Rizzoli [28]):

1. The initial state vector is assumed to be Gaussian with known mean $E\{x_0\}$ and known error covariance matrix $P_0 = Cov\{x_0\}$

2. The model noise w_k is a stationary white noise process with zero mean and known covariance matrix $Q_k = Cov\{w_k\}$

3. The measurement noise v_k is a stationary white noise process with zero mean and known covariance matrix $R_k = Cov\{v_k\}$

4. The initial state x_0, the model noise w_k and the measurement noise v_k are mutually independent

The optimal state estimate and the associated error covariance matrix can be calculated recursively. First, during the forecast step, the optimal state estimate is propagated from time k to time $k + 1$. Moreover the model errors are propagated during the same period using the model dynamics forced by model noise. This step is represented by the following equations:

$$x_{k+1}^f = A_k x_k^a + B_k u_k \tag{3.3}$$

$$P_{k+1}^f = A_k P_k^a A_k^T + \Lambda_k Q_k \Lambda_k^T \tag{3.4}$$

where the superscripts a and f represent analysed or updated values and forecasted or estimated values, respectively. $P_k \epsilon \Re^{n \times n}$ is the error covariance matrix of the estimation and Q_k is the error covariance matrix of the system.

If measurements are available at time $k+1$, the state vector and the error covariance matrix are updated during the analysis step using the following equations:

$$x_{k+1}^a = x_{k+1}^f + K_{k+1}(z_{k+1} - H_{k+1} x_{k+1}^f) \tag{3.5}$$

$$P_{k+1}^a = P_{k+1}^f - K_{k+1} H_{k+1} P_{k+1}^f \tag{3.6}$$

where $K_k \epsilon \Re^{n \times p}$ is the Kalman gain or weighting matrix, which describes the weights given to the innovation vector (difference between observed and forecasted variables) in the calculation of the analysed state variables. The Kalman gain is conventionally obtained from the equation

$$K_{k+1} = P_{k+1}^f H_{k+1}^T [H_{k+1} P_{k+1}^f H_{k+1}^T + R_{k+1}]^{-1} \tag{3.7}$$

with $R_k \epsilon \Re^{p \times p}$ the error covariance matrix of the measurement noise.

Equations 3.3 - 3.7 constitute the Kalman filter algorithm that provides the optimal estimate of the system state in a least square sense when all the aforementioned assumptions are fulfilled. In a real application the assumptions are normally violated by:

- Non-linear system dynamics and/or measurement equation

- Unknown or partly known initial conditions, $E\{x_0\}$ and P_0

- Unknown or partly known noise statistics, Q_k and R_k

The full Kalman filter applied on a large system is extremely expensive to compute due to a very large requirement on numerical operations and storage. The propagation of the error covariance matrix through equation 3.4 requires an amount of computation equivalent to $2n$ times the propagation of the model one time step forward. Moreover, the error covariance matrix, which has a size of $n \times n$ has to be stored.

3.1.1   Sequential updating algorithm

When the data assimilation is based on in-situ measurements which are normally sparsely distributed in space, the measurement matrix H_{k+1} has only a few non-zero elements. In this case, only the columns of the error covariance matrix P_{k+1}^f that correspond to the non-zero elements of H_{k+1}, have to be calculated. To avoid the calculation of the inverse matrix in equation 3.7, the updating of the state vector as well as the error covariance matrix can be carried out sequentially. This approach is especially efficient when the measurement errors are uncorrelated and the matrix R_k becomes diagonal. If this is not the case, a transformation of R_k using an orthogonal matrix T_k has to be carried out, so that $T_k R_k T_k^T$ is a diagonal matrix (see Chui and Chen [11]).

Following Chui and Chen [11] and assuming that R_k is a diagonal matrix, the sequential updating of the state vector and the error covariance matrix are given by

$$x_{k+1,i}^a = x_{k+1,i-1}^a + K_{k+1,i}[z_{k+1,i} - H_{k+1,i}x_{k+1,i-1}^a], \quad x_{k+1,0}^a = x_{k+1}^f \qquad (3.8)$$

$$P_{k+1,i}^a = P_{k+1,i-1}^a - K_{k+1,i}H_{k+1,i}^T P_{k+1,i-1}^a, \quad P_{k+1,0}^a = P_{k+1}^f \qquad (3.9)$$

where $K_{k+1,i}$ is the Kalman gain vector associated to the ith measurement $z_{k+1,i}$, and has been obtained from the equation

$$K_{i,k} = \frac{1}{H_{k+1,i}P_{k+1,i-1}^a H_{k+1,i}^T + \sigma_{mi}^2} P_{k+1,i-1}^a H_{k+1,i}^T \qquad (3.10)$$

Here $H_{k+1,i}$ is the ith column of the matrix H_{k+1}, while σ_{mi} is the standard deviation of the noise of the ith measurement. The sequential algorithm [3.8 - 3.10] is repeated for all p measurements. The state vector and the error covariance matrix, which are obtained after the assimilation of the last measurement, are:

$$x_{k+1}^a = x_{k+1,p}^a, \quad P_{k+1}^a = P_{k+1,p}^a \qquad (3.11)$$

3.1.2   State vector augmentation: time coloured noise

The error propagation equations of the Kalman filter have been developed under the assumption that the system noise is uncorrelated in time. Gelb [27] introduced a procedure which consists of including the correlated noise into the state vector and so at the same time extending the propagation matrix with new rows and columns. The new state variables are considered as part of the linear dynamic system which itself is excited by noise. The time correlation can be described by different correlation models such as random constant, random walk, random ramp and exponentially correlated random variable, all of which are described in Gelb [27]

A new propagation matrix should be defined and consequently the forecast equation and the propagation of the error covariance have to be modified. The system noise w used in equation 3.1 is substituted by the time correlated noise $x^{.,w}$ which can be modelled using the following differential equation

$$x_{k+1}^{f,w} = A_k^w x_k^{a,w} + w_k^w \tag{3.12}$$

The new model propagation equation without forcing is rewritten as:

$$x_{k+1}^{f+} = \begin{pmatrix} x_{k+1}^f \\ x_{k+1}^{f,w} \end{pmatrix} = \begin{pmatrix} A_k & \Lambda_k \\ 0 & A_k^w \end{pmatrix} \begin{pmatrix} x_k^a \\ x_k^{a,w} \end{pmatrix} + \begin{pmatrix} 0 \\ I \end{pmatrix} [w_k^w] \tag{3.13}$$

where w^w is an uncorrelated random noise and A_k^w is the propagator for the specific correlation model considered. Verlaan and Heemink [64] use an auto-regressive AR(1) process as a time correlation model.

3.2 Extended Kalman filter

When the filter is applied to non-linear models (see for example, Jazwinsky [35], Ghil and Malanotte-Rizzoli [28] and Evensen [21]) successive linearisations of the non-linear dynamics have to be carried out. It is this new approach that is known as the extended Kalman filter (EKF). Equation (3.1) is now written as

$$x_{k+1} = f(x_k, u_k) + \Lambda_k w_k \tag{3.14}$$

where $f(x_k, u_k)$ is a non linear vector function describing the system dynamics (Heemink, [31]). The model error, as in the linear case, is assumed to be additive; therefore the forecast of the state vector is given by the deterministic model equation

$$x_{k+1}^f = f(x_k^a, u_k) \tag{3.15}$$

The propagation of the error covariance is based on a Taylor series expansion of $f(x_k, u_k)$ around the current estimate of the state vector (Evensen [22]). This propagation can be written as:

$$P_{k+1}^f = F_k P_k^a F_k^T + \Lambda_k Q_k \Lambda_k^T + F_k \Theta_k \mathcal{H}_k^T + \frac{1}{4} \mathcal{H}_k \Gamma_k \mathcal{H}_k^T$$

$$+\frac{1}{3}F_k\Gamma_k\mathcal{T}_k^T - \frac{1}{4}\mathcal{H}_k P_k^a (P_k^a)^T \mathcal{H}_k^T + \dots \tag{3.16}$$

where the tangent linear operator F_k is defined by the Jacobian matrix

$$F_k = \left.\frac{\partial f(x)}{\partial x}\right|_{x=x_k^f} \tag{3.17}$$

$\mathcal{H}$ is the Hessian and $\mathcal{T}$ contains third order derivatives of $f(x)$. Θ and Γ are third and fourth order statistical moments respectively. By using the transition matrix defined in equation (3.17), the error evolution is correct to first order. For the Extended Kalman filter, higher order moments have been neglected, even though it is accepted that these may became significant when the system is strongly non-linear. The equation 3.16 is then truncated to

$$P_{k+1}^f = F_k P_k^a F_k^T + \Lambda_k Q_k \Lambda_k^T \tag{3.18}$$

Evensen [21] showed that the implementation of the extended Kalman filter for data assimilation in a multilayer quasi-geostrophic (QG) model presents a closure problem in the error covariance evolution equation, implying an unbounded error covariance growth. This is the consequence of discarding higher order moments from the above-mentioned equation. Direct EKF applications can be found in several different scientific fields. Examples of these applications can be found in control theory (Gelb [27] and Jazwinsky [35]), while within oceanography the EKF has been applied to a non-linear quasi geostrophic ocean circulation model (see Evensen [21]). It has been further applied in meteorology (Cohn and Parrish [12]), in hydraulics and environmental engineering on a 2-D advection dispersion model for state correction and parameter estimation (Cañizares [8]) and also within hydrology for forecasting purposes (Wood *et al.* [67]).

3.3   The Reduced Rank SQuare RooT filter

The Reduced Rank SQuare RooT filter (RRSQRT) can be included in the group of suboptimal schemes that uses a simplification in the error covariance modelling. Originally introduced by Verlaan and Heemink [63], the RRSQRT filter uses a square root algorithm as well as a lower-rank approximation of the propagated analysis-error covariance matrix.

The lower-rank approximation reduces the computational burden associated with the use of the Kalman filter by providing a reduced rank factorisation of the error covariance matrix. The use of a matrix of lower rank can fail to preserve the property of positive-semidefinitness of the error covariance matrix, leading to instabilities in the filter (Verlaan [61]). By using a square root algorithm the error-covariance matrix is always positive-semidefinite because no negative eigenvalues can occur. The square root algorithm introduced by Potter [53] and Maybeck [47] makes use of the Cholesky factorization. The square root factorization for the RRSQRT is based on eigendecomposition. Verlaan and Heemink [61] illustrate the factorisation in the following way. Consider the eigendecomposition of the error covariance matrix P as

$$P = VDV^T \tag{3.19}$$

Where the columns of V contain the eigenvectors and D is a diagonal matrix containing the eigenvalues of P. A square root factor of P can be defined as

$$S = VD^{\frac{1}{2}} \tag{3.20}$$

The eigendecomposition can be easily used for approximation by matrices of lower rank. The truncation of the eigendecomposition using the largest q eigenvalues is the optimal rank q approximation in the case of a symmetric positive semidefinite matrix.

3.3.1 The RRSQRT algorithm

As introduced by Verlaan and Heemink [64] and defined in more detail in Verlaan [62], the RRSQRT algorithm can be divided into three main steps: a propagation step, a reduction step and a measurement update, as follows.

Propagation step

During the propagation step, the state vector as well as the error covariance matrix are propagated one time step forward. The propagation of the state vector is performed through equation 3.15 i.e.

$$x_{k+1}^f = f(x_k^a, u_k) \tag{3.21}$$

The propagation of the error covariance matrix (originally in the extended Kalman filter equation 3.18) is carried out now as

$$S_{k+1}^{*f} = \left[F_k S_k^a \ \middle| \ \Lambda_k Q_k^{1/2} \right] \tag{3.22}$$

where the matrix $S_k^a \epsilon \Re^{n \times q}$ represents the approximation of rank q of the square root of the error covariance matrix P_k^a at time step k, with q the number of leading eigenvalues. This matrix has only q columns, while the complete error covariance matrix has n. It is important to notice that n is much larger than q ($n \gg q$). $Q_k^{1/2}$

represents the square root of the error covariance matrix of the system noise. The expression on the right side of equation (3.22) denotes a matrix built with two block matrices; thus matrix S_{k+1}^{*f} has a dimension of $q + m$.

To avoid the calculation of the first-order derivatives to obtain F_k, Verlaan and Heemink [64] used a finite difference approximation of the tangent linear operator. Introducing $s_{j,k}^a$ as the jth column of the matrix S_k^a, then

$$F_k S_k^a = \frac{\partial f}{\partial x} S_k^a = \frac{\partial f}{\partial x} \left[s_{1,k}^a, \ldots, s_{q,k}^a \right] =$$
$$\left[\frac{\partial f}{\partial x} s_{1,k}^a, \ldots, \frac{\partial f}{\partial x} s_{q,k}^a \right] \qquad (3.23)$$

Every vector column can be approximated using a first order finite difference approximation such as:

$$\frac{\partial f}{\partial x} s_{j,k}^a \approx \frac{f(x_k^a + \epsilon s_{j,k}^a, u_k) - f(x_k^a, u_k)}{\epsilon} \qquad (3.24)$$

where ϵ is a small value. This approach may fail in the same way as the extended Kalman filter if the model presents strong non-linearities or discontinuities.

Reduction step

The aim of this step is to reduce the number of columns of the square root approximation of the error covariance matrix S_{k+1}^{*f} from $q + m$ to q in order to maintain a constant number of columns equal to q throughout the assimilation process. The best approximation will be obtained by using only the leading eigenvalues and eigenvectors of the approximated error covariance matrix $P_{k+1}^{*f} = S_{k+1}^{*f} [S_{k+1}^{*f}]^T$. By computing the eigendecomposition of $[S_{k+1}^{*f}]^T S_{k+1}^{*f}$ it is possible to obtain a more efficient procedure than computing the eigendecomposition of P_{k+1}^{*f} or the singular value decomposition of S_{k+1}^{*f} (Verlaan [61]). The procedure can be summarised as follows:

The eigendecomposition of the matrix $[S_{k+1}^{*f}]^T S_{k+1}^{*f}$ is:

$$[S_{k+1}^{*f}]^T S_{k+1}^{*f} = V_{k+1} D_{k+1} V_{k+1}^T \qquad (3.25)$$

where $V_{k+1} \epsilon \Re^{(q+m) \times (q+m)}$ is the matrix of the orthonormal eigenvectors and the diagonal matrix containing the eigenvalues is represented by $D_{k+1} \epsilon \Re^{(q+m) \times (q+m)}$.

It is shown in Verlaan [61] that

$$P_{k+1}^{*f} = \left(S_{k+1}^{*f} V_{k+1} D_{k+1}^{-1/2}\right) D_{k+1} \left(S_{k+1}^{*f} V_{k+1} D_{k+1}^{-1/2}\right)^T \tag{3.26}$$

is the eigendecomposition of P_{k+1}^{*f} and thus $S_{k+1}^{f} \epsilon \Re^{n \times q}$ is the square root of the optimal rank q decomposition of P_{k+1}^{*f}. It is given by the expression

$$S_{k+1}^f = \begin{bmatrix} S_{k+1}^{*f} & V_{k+1} \end{bmatrix}_{1:n, 1:q} \tag{3.27}$$

Measurement update

In Section 3.1.1 the sequential measurement update used in the Kalman filter was presented. Maybeck [47] introduced a similar procedure (based on Potter [53]) to assimilate scalar measurements and vector measurements into a square root Kalman filter. The procedure is composed from the following equations

$$a_{k+1,i} = [S_{k+1,i-1}^a]^T H_{k+1,i}^T \tag{3.28}$$

$$\gamma_{k+1,i} = \frac{1}{[a_{k+1,i}]^T a_{k+1,i} + \sigma_{mi}^2} \tag{3.29}$$

$$K_{k+1,i} = S_{k+1,i-1}^a a_{k+1,i} \gamma_{k+1,i} \tag{3.30}$$

$$S_{k+1,i}^a = S_{k+1,i-1}^a - K_{k+1,i} a_{k+1,i}^T \frac{1}{1 + \sqrt{\gamma_{k+1,i} \sigma_{mi}^2}}, \quad S_{k+1,0}^a = S_{k+1}^f \tag{3.31}$$

$$x_{k+1,i}^a = x_{k+1,i-1}^a + K_{k+1,i}[z_{k+1,i} - H_{k+1,i} x_{k+1}^f], \quad x_{k+1,0}^a = x_{k+1}^f \tag{3.32}$$

where the notation is the same as it was in Section 3.1.1. After the assimilation of the the last measurement (the pth) the analysed state vector and the analysed square root approximation of the error covariance matrix are obtained as.

$$x_{k+1}^a = x_{k+1,p}^a, \quad S_{k+1}^a = S_{k+1,p}^a \tag{3.33}$$

3.4   The Ensemble Kalman filter

The Ensemble Kalman filter (EnKF) introduced by Evensen [22] is based on Monte Carlo simulations. The EnKF is a sequential data assimilation method where the error statistics are predicted using Monte Carlo or ensemble integration (Evensen

[24]). An ensemble of model states is integrated forward in time. From the ensemble, all the statistical information together with statistical moments (such as mean and covariance) can be calculated. The EnKF presents two important advantages as compared to the extended Kalman filter: it solves the closure problem presented in Evensen [21] and reduces drastically the computational load and the storage requirement. The computational load of the EnKF corresponds to M model integrations, with M the ensemble size, while the storage requirement consists of M model state vectors of size n. Evensen [24] claims that for practical ensembles sizes $\mathcal{O}(100)$ the errors will be dominated by statistical noise, not by closure problems or unbounded error variance growth. Another feature of the EnKF, pointed out by Madsen [45] is that while in the Kalman filter the model errors are additive and explicitly related to the state vector, in the EnKF the model errors can be directly related to the elements in the numerical model that are expected to be poorly described.

3.4.1   The Ensemble Kalman filter scheme

Following Evensen [24] and Madsen [45], a description of the EnKF scheme is provided here for future reference.

An ensemble of size M of initial states $x_{j,0}^a$ with $j = 1, \ldots, M$ is generated. The ensemble mean is an estimate of the best guess initial condition $E\{x_0\}$ and the ensemble covariance represents the uncertainty $P_0 = Cov\{x_0\}$ in the first guess initial state.

Each ensemble member is propagated forward in time and forced by model errors $w_{j,k}$, as described by the following equation:

$$x_{j,k+1}^f = f(x_{j,k}^a, u_k, w_{j,k}) \qquad \text{for} \quad j = 1, \ldots, M \tag{3.34}$$

The addition of these errors has the same effect as adding the system error covariance matrix in the Kalman filter (for example equation 3.18). The model error $w_{j,k}$ is randomly drawn from a predefined distribution with zero mean and covariance matrix Q_k.

Any quantile of the ensemble forecast can be used as the forecast of the state vector. Evensen [22] uses the mean value, which is calculated as

$$\bar{x}_{k+1}^f = \frac{1}{M} \sum_{j=1}^{M} x_{j,k+1}^f \tag{3.35}$$

Moreover, the error covariance matrix of the forecast is obtained from the ensemble as

$$P^f_{k+1} = S^f_{k+1}[S^f_{k+1}]^T \qquad s^f_{j,k+1} = \frac{1}{\sqrt{M-1}}(x^f_{j,k+1} - \bar{x}^f_{k+1}) \qquad (3.36)$$

where $s^f_{j,k+1}$ is the jth column of S^f_{k+1}.

As introduced by Burgers *et al.* [7] it is essential to add random perturbations to the measurements otherwise the assumption of measurements being random variables will not be valid. When these perturbations are not added to the observations, the analysed covariance will have a very small variance. Therefore an ensemble of size M of possible observations is generated

$$z_{j,k+1} = z_{k+1} + v_{k+1} \qquad \text{for} \quad j = 1, \ldots, M \qquad (3.37)$$

where z_{k+1} is the actual measurement vector and $v_{j,k+1}$ is the measurement error randomly generated from a distribution with zero mean and covariance matrix R_{k+1}

The Kalman gain is calculated from the same equation as for the extended Kalman filter, i.e. equation 3.7, using the forecast error covariance matrix P^f_{k+1}

Each model state ensemble member is updated using a similar expression to that used in the standard Kalman filter:

$$x^a_{j,k+1} = x^f_{j,k+1} + K_{k+1}(z_{j,k+1} - H_{k+1}x^f_{j,k+1}) \qquad \text{for} \quad j = 1, \ldots, M \qquad (3.38)$$

Finally the updated state vector and the error covariance matrix are obtained from the updated ensemble

$$\bar{x}^a_{k+1} = \frac{1}{M}\sum_{j=1}^{M} x^a_{j,k+1} \qquad (3.39)$$

$$P^a_{k+1} = S^a_{k+1}[S^a_{k+1}]^T \qquad S^a_{j,k+1} = \frac{1}{\sqrt{M-1}}(x^a_{j,k+1} - \bar{x}^a_{k+1}) \qquad (3.40)$$

Under the assumption of uncorrelated measurement errors a sequential updating algorithm can be applied in the EnKF. This assumption is usually valid when in-situ data are assimilated. Using this algorithm, it is not neccesary to compute and store the full error covariance matrix P_{k+1}. The sequential updating can be carried out using the matrix $S^f_{k+1} \epsilon \Re^{n \times M}$ with $n \gg M$. The algorithm is almost the same as the one presented in section 3.3.1 for the measurement update of the RRSQRT. The differences in the case of the EnKF are that the update is now carried out for every ensemble member and the updated error covariance matrix is calculated from the updated ensemble using equation 3.40 instead of equation 3.31.

Evensen and van Leeuwen [25] dealt with the problem of spatially distributed data. In this case, if the number of data is larger than the ensemble size, the matrix to be inverted in equation 3.7 may be ill-conditioned. These authors also proposed a data reduction scheme to reduce the dimension of the matrix to be inverted together with an updating scheme.

3.5 Integration of a sequential data assimilation technique in MIKE 21

After the general and theoretical description of the Kalman-filter-based algorithms, special considerations have to be taken into account for integrating this with a specific model and for a specific application. A description of variables, errors, observations and some particular aspects of the implementation are provided in this section.

3.5.1 Definition of the state vector

The general definition of the state vector in MIKE 21 has been already described in chapter 2. The variables included in the state vector are surface elevations ζ and flux densities p at time step k and the flux densities q at time steps $k + \frac{1}{2}$ and $k - \frac{1}{2}$. The direct use of these variables in the RRSQRT filter produces the following problem. The order of magnitude of water levels is small compared with the order of magnitude of the flux densities, and hence in the error covariance matrix the error associated with the flux densities is much larger than the error associated with the water levels. The leading eigenvalues of the error covariance matrix will be associated with the error in the flux densities and never with the water levels. In order to overcome this problem depth average velocities in x and y directions, which have similar order of magnitude to water levels, are used instead of flux densities. These variables are related with the water depth h by the expression

$$p_k = \mathrm{v}_{x,k} h_k \quad , \quad q_{k+\frac{1}{2}} = \mathrm{v}_{y,k+\frac{1}{2}} h_{k+\frac{1}{2}} \quad , \quad q_{k-\frac{1}{2}} = \mathrm{v}_{y,k-\frac{1}{2}} h_{k-\frac{1}{2}} \quad (3.41)$$

where v_x and v_y are the components of the depth averaged velocity in the x and y directions. Once again either a rectangular velocity distribution is assumed, or a depth-averaged velocity. The state vector can then be written as

$$x_k = \begin{pmatrix} \zeta_k \\ \mathrm{v}_{x,k} \\ \mathrm{v}_{y,k+\frac{1}{2}} \\ \mathrm{v}_{y,k-\frac{1}{2}} \end{pmatrix} \quad (3.42)$$

Further explanations about the scaling problem are included at the end of the present chapter.

3.5.2 Measurements and measurement noise definition

The sequential updating algorithm, which has been introduced during this chapter, was used in this implementation. No spatially distributed data (altimetry or satellite data) were assimilated in the present case.

Direct measurements of the state vector variables are assimilated. They consist of surface elevation and velocity components observed at stations near the coast or at offshore platforms and spatially distributed throughout the model. The measurements are usually obtained at different time intervals than the model time steps, and therefore they have to be transformed in order to coincide temporally with the model time steps.

It is also assumed that one observation is associated with only one grid point, i.e. one state variable, so the vector $H_{k+1,i}$ (introduced in section 3.1.1) which contains the relations between the ith measurement and the model variables, has all the elements zero except the one associated with the observed variable that is equal to 1. This assumption is, probably, one of the main sources of the measurement noise. It is difficult to quantify how representative a point measurement is for the entire grid. Moreover, the measurement noise is also associated with the measurement devices accuracy.

Another further assumption is that the measurement errors are independent of the model errors, while being also mutually uncorrelated and homogeneous in time. Under these assumptions, the sequential updating algorithm, introduced in previous sections, can be used.

3.5.3 The definition of the system noise

The definition of the system noise in the filter seems to be one of the most influential factors for a good filter performance. Usually, the system noise is unknown, not only in terms of the source but also in terms of the magnitude of the variance (the mean is assumed equal to zero because this is a white noise) and the spatial correlation. Regarding the source of the noise, three different categories should be considered:

1. Errors in the forcing terms. These errors are related to the prescribed open boundary conditions and the meteorological forcing terms. Despite the property that the tidal component of the surface elevation at the open boundary can be obtained quite accurately, the variations due to the meteorological effects, both within the boundary and from outside the boundary of the modelled domain, often remain uncertain. Moreover, the wind field as well as the pressure gradient field contains errors.

2. Errors in the model parameters. Some of the model parameters cannot be measured directly in nature and therefore they must be estimated implicitly,

based on the model dynamics. In the case of the parameters that can be measured, the measurements available are normally not sufficient to provide a good estimate of the parameters for the entire model domain. Among these parameters are the wind drag coefficient and the bottom friction. Another spatially distributed model parameter that can often produce errors in the model is the use of an incorrect bathymetry.

3. Errors in the model description. These errors are related to poorly-described or neglected physical processes in the system equation and mathematical approximations including the errors generated at unresolved or sub-grid-scale (SGS) motions (McWilliams [48] and Pinardi [52]).

The most usual way of quantifying the covariance matrix of the system noise Q_k is by using the expression:

$$Q_k^{i,j} = \sigma_i \sigma_j \rho(l) \tag{3.43}$$

where:

$Q_k^{i,j}$ the covariance of the system noise between grid points i and j
σ_i, σ_j the standard deviation of the system noise at grid points i and j
l distance between points i and j
$\rho(l)$ the correlation coefficient of the system noise at distance l

Different correlation functions can be used to define the system noise correlation structure. Zhou [69] and Marsily [46], present different correlation functions, such as exponential, Gaussian, spherical and Inverse distance, which are commonly used in Kalman filter applications. An introduction to these correlation functions is given here:

The exponential correlation model can be defined by

$$\rho(l) = e^{(l/b)} \tag{3.44}$$

where b indicates the magnitude of the spatial correlation. l_c is the spatial correlation length, which is defined as the average distance at which the system noise is correlated. In the exponential correlation model, $l_c = b$.
The Gaussian correlation model is defined by

$$\rho(l) = e^{[-(l/b)^2]} \tag{3.45}$$

where again b indicates the magnitude of the spatial correlation. The correlation length for the Gaussian model is given by

$$l_c = b\sqrt{\pi/2} \tag{3.46}$$

The Spherical correlation model is defined by

$$\rho(l) = 1 - 1.5(l/\lambda) + 0.5(l/\lambda)^3 \quad l < \lambda$$
$$\rho(l) = 0 \quad l > \lambda \tag{3.47}$$

where λ is the correlation range. The correlation length for the Spherical model is

$$l_c = \frac{3}{8}\lambda \tag{3.48}$$

A special feature of the Spherical model is that there is no correlation beyond the correlation range

The Inverse distance correlation model is defined by

$$\rho(l) = \rho^l \tag{3.49}$$

where ρ is the correlation coefficient. When $\rho = e^{\frac{-1}{b}}$ the exponential and the inverse distance models are identical. The correlation length for the Inverse distance model is

$$l_c = \frac{1}{\ln(\rho)} \tag{3.50}$$

When more than one error source is considered, the correlations among the errors have to be defined. Typically, it has been assumed that the most influential errors are the errors in the forcing terms, which are included in the first group. Moreover, it is also commonly assumed that errors in the open boundaries and forcing terms are uncorrelated.

On the definition of the system noise in a coarser grid

Since one of the most important source of errors in the model is in the meteorological forcing input, and because this forcing is very often obtained from a coarse grid, the system noise can be defined on a grid coarser than the model grid (Heemink [31]). This approach has two main advantages. First, the number of noise points m is smaller, which is very important in a RRSQRT filter application because the eigenvalue decomposition of a matrix of order $(q + m \times q + m)$ has to be calculated. The second advantage, as established by Heemink, is that the energy of the short waves introduced in the model grid by the filter is limited.

The interpolation matrix Λ_k can be calculated using a Kriging technique. Kriging is a statistical optimal interpolator. As mentioned in Marsily [46] it is an exact interpolator because it gives no uncertainty at a measured point. It is the best linear unbiased estimator of a spatial variable at a particular site. Kriging assigns low weights to distant samples and vice versa, but also takes into account the relative position of the samples to each other and to the site of the area being estimated.

In our case, the aim is to interpolate the noise that has been defined on a coarse grid into the model grid. Because the noise statistics are given and they are invariant in time, the process is second-order stationary with known mean and covariance. Simple Kriging can be used here.

The values of the variable $z_0(p)$ are known at the n_{g0} points, $p_1, p_2, \ldots, p_{n_{g0}}$ of the grid $G0$. The problem is to estimate a quantity z_1 in the model grid $G1$. To estimate

z_1 we consider a weighted average of the n_{g0} available data:

$$z_1 = \sum_{i=1}^{n_{g0}} \lambda_i z_0(p_i) \qquad (3.51)$$

where λ_i are the weights of the Kriging estimator. These weights are the unknowns of the problem because they specify the relationship between one point of $G1$ with points in $G0$. The weights have to be calculated under two assumptions:

1. The estimator has to be unbiased, so the mathematical expectations of the estimation z_1^* and the true value z_1 have to be equal.

$$E(z_1^* - z_1) = 0 \qquad or \qquad E(z_1^*) = E(z_1) \qquad or \qquad \sum_{i=1}^{n_{g0}} \lambda_i = 1 \qquad (3.52)$$

2. The error estimation has to be minimal, i.e. the variance of the error in the estimation has to be minimum

$$E\left[(z_1^* - z_1)^2\right] \qquad minimum \qquad (3.53)$$

As described in Marsily [46] and Delhomme [18] the problem consists of minimising a quadratic form of the unknowns λ_i. The minimisation of equation 3.53 subject to the linear constraint in equation 3.52, can be found by minimising the quantity

$$\frac{1}{2}E\left[(z_1^* - z_1)^2\right] - \mu\left[\sum_{i=1}^{n_{g0}} \lambda_i - 1\right] \qquad (3.54)$$

where μ is unknown and is called a Lagrange multiplier. The minimum of the quadratic form in λ and μ is obtained by setting to zero its partial derivatives with respect to λ_i and μ. The following linear system, called the Kriging system, is then obtained:

$$\sum_{j=1}^{n_{g0}} \lambda_j \gamma(z_0(p_i) - z_0(p_j)) + \mu = \gamma(z_0(p_i) - z_1) \qquad i = 1, \ldots, n_{g0}$$

$$\sum_{i=1}^{n_{g0}} \lambda_i = 1 \qquad (3.55)$$

where the variograms γ are given by the expressions:

$$\gamma_{i,j} = \gamma(z_0(p_i) - z_0(p_j)) = \frac{1}{2}E\left[(z_0(p_i) - z_0(p_j))^2\right]$$

$$\gamma_{i,z_1} = \gamma(z_0(p_i) - z_1) = \frac{1}{2}E\left[(z_0(p_i) - z_1)^2\right] \qquad (3.56)$$

If we defined $\gamma(l)$ as the variogram or mean quadratic increment between two points separated by the distance l and suppose that the process is second order stationary, the variogram could be defined in terms of the previously-defined covariance function (equation 3.43) in the following way:

$$\gamma(l) = \sigma_0 \sigma_0 \rho(0) - \sigma_0 \sigma_l \rho(l) \tag{3.57}$$

where σ is the standard deviation and ρ the correlation model. The two terms of equation 3.57 represent the variance at point 0 and the covariance for a point a distance l from point 0. Assuming that the standard deviation σ is the same at every point, equation 3.57 can be written as

$$\gamma(l) = \sigma^2 (1 - \rho(l)) \tag{3.58}$$

For every point of $G1$ the complete linear system has to be calculated. The Kriging system in linear form is

$$
\begin{pmatrix}
0 & \gamma_{1,2} & \gamma_{1,3} & \cdots & \gamma_{1,n_{g0}} & 1 \\
\gamma_{2,1} & 0 & \gamma_{2,3} & \cdots & \gamma_{2,n_{g0}} & 1 \\
\gamma_{3,1} & \gamma_{3,2} & 0 & \cdots & \gamma_{3,n_{g0}} & 1 \\
\vdots & \vdots & \vdots & \ddots & \vdots & \vdots \\
\gamma_{n_{g0},1} & \gamma_{n_{g0},2} & \gamma_{n_{g0},3} & \cdots & 0 & 1 \\
1 & 1 & 1 & \cdots & 1 & 0
\end{pmatrix}
\begin{pmatrix}
\lambda_1 \\
\lambda_2 \\
\lambda_3 \\
\vdots \\
\lambda_{n_{g0}} \\
\mu
\end{pmatrix}
=
\begin{pmatrix}
\gamma_{1,z_1} \\
\gamma_{2,z_1} \\
\gamma_{3,z_1} \\
\vdots \\
\gamma_{n_{g0},z_1} \\
1
\end{pmatrix}
\tag{3.59}
$$

The matrix of the Kriging equations only needs to be inverted once for all the points z_1 in $G1$.

Errors at the open boundaries

In the cases where only noise in the boundary conditions is considered, additive noise, as in equation 3.15, can be used.

$$x_{k+1}^f = f(x_k^a, u_k) + \Lambda_k w_k \tag{3.60}$$

In the typical case, where the boundary is defined as a water level boundary, the noise is associated directly with the variable water level which is included into the state vector. The noise could be defined using the correlation models previously introduced, and using a coarser grid. It should be noticed here that in this case the boundary is defined along a line and not spatially distributed. Moreover, when more than one boundary is defined in the model, the assumption of absence of correlation between errors at the boundaries is usual; this assumption is commonly used in North Sea models, even though it is not always entirely correct.

Error in the meteorological forcing terms

The initial step in the definition of the error in the meteorological forcing is to defined the noise on a grid which is coarser than the model grid as it was defined in section 3.5.3. Values from the coarser grid are interpolated into the finer grid using for example a Kriging technique where the statistics used are the ones associated with the error covariance matrix of the system Q.

In the Kalman filter and subsequently the RRSQRT filter, the model errors are explicitly related through the state vector and they are assumed to be additive. On the other hand, in the EnKF the model errors can be directly related to any element in the numerical model.

To deal with the aforementioned limitation, two possibilities were considered in Cañizares [9]. The first and the simplest way to define the error is by using additive noise and by associating it directly with the average water-velocity components. In this case, the error covariance matrix of the system noise Q comprehends the errors associated with the average water velocity components which have been introduced through the incorrect wind input. This assumption, as well as the usual definition of the applicability of a homogeneously distributed error throughout the model, can lead to instabilities due to unrealistic or physically contradictory filter corrections (Abbott and Minns [2]). In fact, the field is far from homogeneous in a real-world application and indeed the values of the velocities may vary over a wide range. Once the standard deviation of the noise has been selected, this value can represent a very small number compared with the velocities in some locations, while at other points it can be very large. At the points of the first group the filter corrects too little while in points of the second group the correction can be too large and divergences are sure to occur. With the aim of reducing this effect, the error can be scaled as a function of the bathymetry. Deep areas present, normally, smaller depth-averaged velocities than shallow areas: therefore the noise in the deeper areas should be smaller than in the shallower areas.

The second approach consists of relating directly the error to the meteorological forcing terms. These terms are included in the momentum equations in the x- and y-directions. The momentum equation in the x-direction in finite difference form can be expressed as;

$$AMO_j \zeta_j^{n+\frac{1}{2}} + BMO_j p_j^{n+1} + CMO_j \zeta_{j+1}^{n+\frac{1}{2}} = DMO_j \qquad (3.61)$$

where the meteorological forcing terms (the wind friction term and the atmospheric pressure variation term) are included in the coefficient DMO. In the momentum equation in the x-direction, as introduced earlier in equation 2.2, these terms are

$$fVV_x \quad ; \quad \frac{h}{\rho_w} \frac{\partial}{\partial x}(P_a) \qquad (3.62)$$

Similar expressions to equations 3.61 and 3.62 can be written for the momentum equation in the y-direction. The meteorological forcing terms are included in the coefficient DMO in a finite difference form. In this case the system noise is initially defined in the matrix Q_k as the error in the coefficient DMO. The Kalman filter equations for the state and the error covariance propagation 3.60 and 3.18 can now be written as

$$x^f_{k+1} = f(x^a_k, u_k + \Lambda_k w_k) \tag{3.63}$$

$$P^f_k = F_k P^a_{k-1} F^T_k + G_k \Lambda_k Q_k \Lambda^T_k G^T_k \tag{3.64}$$

with

$$G_k = \left. \frac{\partial f(x)}{\partial u} \right|_{x=x^f_k; u=u_k} \tag{3.65}$$

The operator G_k can be solved using a finite difference approximation as in the case of the tangent linear operator F_k. The propagation equation of the square root approximation of the error covariance matrix in the case of the RRSQRT can now be written as,

$$S^f_{k+1} = \left[F_k S^a_k \quad \bigg| \quad G_k \Lambda_k Q^{\frac{1}{2}}_k \right] \tag{3.66}$$

Therefore, in order to obtain the propagated error covariance matrix S, the equivalent of q (the number of leading eigenvalues) plus m (the number of total noise input points) model simulations are needed. Defining m_x and m_y as the number of system noise points for DMO in the x and y directions respectively, the term at the right hand side of equation 3.66 can be solved by using the following expression

$$
\begin{aligned}
\frac{\partial f}{\partial u} \Lambda_k Q^{\frac{1}{2}}_k \approx \Big[&f(x_k, u_k + (\Lambda_k Q^{\frac{1}{2}}_k)_1) - f(x_k, u_k), \dots, \\
&f(x_k, u_k + (\Lambda_k Q^{\frac{1}{2}}_k)_{m_x}) - f(x_k, u_k), \\
&f(x_k, u_k + (\Lambda_k Q^{\frac{1}{2}}_k)_{m_x+1}) - f(x_k, u_k), \dots, \\
&f(x_k, u_k + (\Lambda_k Q^{\frac{1}{2}}_k)_{m_x+m_y}) - f(x_k, u_k) \Big]
\end{aligned}
\tag{3.67}
$$

where $(\Lambda_k Q^{\frac{1}{2}}_k)_i$ represents the ith column of the matrix $(\Lambda_k Q^{\frac{1}{2}}_k)$. Under the assumption that the noises in the two directions are uncorrelated and $m_x = m_y$, a simplified procedure for calculating the same term can be written as

$$\frac{\partial f}{\partial u} \Lambda_k Q_k^{\frac{1}{2}} \approx \left[f(x_k, u_k + (\Lambda_k Q_k^{\frac{1}{2}})_1 + (\Lambda_k Q_k^{\frac{1}{2}})_{m_x+1})) - f(x_k, u_k), \ldots, \right.$$

$$\left. f(x_k, u_k + (\Lambda_k Q_k^{\frac{1}{2}})_{m_x} + (\Lambda_k Q_k^{\frac{1}{2}})_{m_x+m_y})) - f(x_k, u_k) \right] \qquad (3.68)$$

In this case only the equivalent of m_x runs of the model are needed to calculate $G_k \Lambda_k Q_k^{\frac{1}{2}}$ instead of the $m_x + m_y$ required with equation 3.67. Moreover, the result matrix has smaller dimensions ($n \times m_x$ instead of $n \times m$) using expression 3.68, thus reducing at the same time the dimension of the eigendecomposition problem.

It is of major importance at this point to define the model in a coarser grid using the interpolation procedure introduced in section 3.5.3 in order to use the smallest value of m_x and m_y as possible.

A different approach has to be used in the case of time-coloured noise (as introduced in section 3.1.2). In this case the propagation equation of the square root approximation of the error covariance is derived using the following equation

$$S_{k+1}^{*f} = \begin{pmatrix} s_{k+1}^{*f} \\ s_{k+1}^{*f,w} \end{pmatrix} = \left[\begin{pmatrix} F_k & G_k \Lambda_k \\ 0 & A_k^w \end{pmatrix} \begin{pmatrix} s_k^a \\ s_k^{a,w} \end{pmatrix} \middle| \begin{pmatrix} 0 \\ Q_k^{1/2} \end{pmatrix} \right] \qquad (3.69)$$

where the matrices with superscript w represent the part of the error covariance related to the augmented variables in the state vector. In this case the finite difference approximation used is:

$$F_k s_k^a + G_k \Lambda_k s_k^{a,w} \approx \left[f(x_k + (s_k^a)_1, u_k + (\Lambda_k s_k^{a,w})_1 - f(x_k, u_k), \ldots, \right.$$

$$\left. f(x_k + (s_k^a)_q, u_k + (\Lambda_k s_k^{a,w})_q - f(x_k, u_k) \right] \qquad (3.70)$$

This is a first order approximation with second order truncation error. An important advantage of this approximation is that it only needs a computation equivalent to q model runs, with q the number of modes.

3.5.4 Scaling problem for a multivariate state vector

In section 3.5.1 was already introduced the problem created by the different scales of magnitude for the different types of variables in the state vector. By augmenting the state vector with new variables such as the DMO error variables, this problem becomes more significant. The eigenvalue decomposition of the covariance matrix does not have to be sensitive to the units of each field. Lermusiaux [42], based on ideas from Preisendorfer [54] indicated two ways to overcome this problem. The first is to divide all the values of each field by a constant. This solution is not always adequate because small errors in some variables can generate large errors in the system state, while other variables can behave completely differently. The second is to introduce normalisation, which will make the errors for every type of variable comparable in magnitude. It consists of dividing the error associated with each variable by the volume and averaged error energy.

The eigenvalue decomposition is calculated on a normalised error covariance matrix. The error covariance matrix S_{k+1}^{*f} obtained from expression 3.69 is divided by the diagonal matrix $N \epsilon \Re^{(n+m)\times(n+m)}$ which has, for each variable, constant elements equal to

$$n_{var} = \sqrt{\frac{\sum_{l=i+1}^{l+n^{var}} \left[S_{k+1}^{*f}\right]_l^T \left[S_{k+1}^{*f}\right]_l}{n^{var}(q+m)}} \qquad (3.71)$$

where n^{var} is the number of grid points associated with the variable n_{var}, index i is the pointer of variable in the state vector and index l represents the lth row of the matrix.

After the eigenvalue decomposition, the truncated error covariance matrix is denormalised by multiplying by the matrix N.

3.5.5 Memory reduction factor

In certain cases during a RRSQRT application, and due to the reduction step, a decay in the magnitude of the error covariance is experienced. This effect occurs when the eigenvalues which represent the old noise are much larger than those corresponding to the new noise and therefore the new noise is lost during the reduction step. This decay in the error covariance produces a consequent deterioration for the filter corrections. The so-called memory reduction factor β is applied on the old term of the error covariance matrix. Its application in, for example, the propagation equation 3.69 leads to the following equation,

$$S_{k+1}^{*f} = \left[\left(\begin{array}{cc} F_k & G_k\Lambda_k \\ 0 & A_k^w \end{array}\right) \beta \left(\begin{array}{c} s_k^a \\ s_k^{a,w} \end{array}\right) \Big| \left(\begin{array}{c} 0 \\ Q_k^{1/2} \end{array}\right)\right] \qquad (3.72)$$

Using a memory reduction factor less than one, the old modes are reduced. As a result the new modes have more influence on the error covariance matrix. When a very small value of β is used, the error covariance matrix remains very small too, yielding to a poor filter performance.

If the open boundaries of the model are not defined as non-reflective boundaries, this problem is likely to appear. In this case the error modes cannot leave the model and remain in it. In Cañizares [9] and for the application of the RRSQRT filter the use of a memory reduction factor was considered.

Chapter 4

Study and comparison: the RRSQRT filter and the EnKF

The objective of the present study is to compare and to test the application of two different data assimilation techniques on a 2D shallow water equation model under different conditions. The main differences between the filters (Madsen [44]) are the following:

- Calculation of the model forecast. The RRSQRT uses a deterministic model forecast while the EnKF calculates the model forecast as the mean (or median, mode) of the ensemble of states.

- Error propagation. The EnKF propagates the ensemble using the full model dynamics while the RRSQRT propagates the error covariance matrix using a tangent linear model operator.

- The model error forcing is introduced in the EnKF as part of the ensemble propagation. In the RRSQRT some matrix algebra is needed for this purpose.

- With regards to the representation of the error covariance matrix, the RRSQRT uses a reduced rank approximation of the square root of the error covariance matrix and the EnKF uses an ensemble estimate of the error covariance.

- The EnKF has a computational cost expressed in terms of the number of model integrations of the order of the ensemble size. In the case of the RRSQRT the greatest part of the computational cost is taken up by a number of model integrations equal to the number of leading eigenvalues plus the cost associated with the matrix algebra including the eigenvalue decomposition, which can be very costly.

In a real application the system noise statistical parameters are unknown or only partly known, so it is very important to learn about the robustness of the filter when it is provided with incorrect or otherwise inappropriate noise statistics. Furthermore, the number of leading eigenvalues in the case of the RRSQRT or the number of ensemble members in the case of the EnKF are very important parameters with respect to the filter performance and the computing and storage loads.

For this purpose a synthetic test (twin test) has been used. The following sections describe the test layout and the results of the first test, Test 1, considering errors in the open boundaries, the second test, Test 2, that includes errors in the meteo- rological forcing terms. Finally some remarks about the mass conservation problem in the Kalman filter are presented.

4.1   Definition of the true model and the observations

The model is defined on a hypothetical bay with a grid of 21×20 points and a grid size of $10 \times 10 \ km^2$. The model has an open boundary at the Northern border. The bathymetry varies from depth values of 100 meters at the centre of the open bound- ary and centre of the bay to 20 meters depth at areas near the closed boundaries. The Chezy bed friction coefficient also varies with the depth, representing values between 30 and 45 $m^{\frac{1}{2}}/s$. Figure 4.1 shows the model bathymetry.

The model is driven by a sinusoidal variation of the surface elevation at the open boundary with a period of 12 hours and a range of 2 meters. In addition a me- teorological forcing is included using wind and pressure fields from an artificially generated moving cyclone that moves in a west-to-east direction with a speed of 8.33 km/h. Figure 4.1 shows the generated cyclonic wind and the pressure fields after 30 hours. The simulation time is set to 48 hours with a time step of 15 minutes. We shall refer to this model as the 'true' model

The synthetic measurements are extracted from the output of the true model and corrupted by adding noise. These measurements consist of time series of the surface elevation at locations (1,16), (8,1) and (20,12). The original time series are disturbed by adding errors that are randomly generated from a Gaussian distribution with zero mean, a diagonal error covariance matrix R_k and the same standard deviation for all the measurements $\sigma_{mi} = 0.05 \ m.$, i.e the measurements are assumed to be independent.

4.2   TEST 1: Errors in the open boundary conditions

The true model has also been distorted by adding errors to the water levels at the open boundary at every time step. The errors are generated from a first order

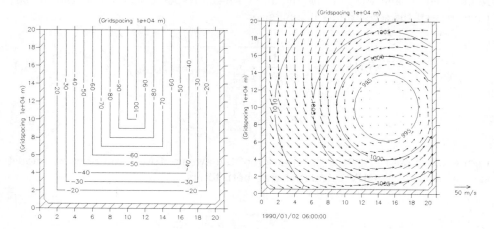

Figure 4.1: Bathymetry (depth in meters) and pressure(Pa) and wind velocity (m/s) fields after 30 hours

autoregressive process given by the expression

$$w_{k+1} = \alpha w_k + \delta_k \tag{4.1}$$

where the lag-one autocorrelation coefficient α is set to 0.9 and the residual errors δ_k are generated from a Gaussian distribution with zero mean and covariance matrix Q_w given by the expression:

$$Q_w^{i,j} = \sigma_i \sigma_j \rho^l \tag{4.2}$$

where an exponential distance correlation model has been used with correlation coefficient $\rho = 0.9$ and l the distance between points i and j in grid co-ordinates. The standard deviation σ is constant and equal to 0.1 m. The state obtained from this simulation is referred to as the "wrong state", and is used as the model forecast.

First, the two filters are compared with respect to the error covariance estimation when no measurements are assimilated. The filter is provided with the noise statistics used to generate the 'wrong' model. The error is then propagated using the model dynamics. The error covariance represents the estimation by the filter of the error in the system. The filter has to be able to estimate correctly the error in the system if a good error reduction is expected. The real error in the system can be obtained as the difference between the 'true' and the 'wrong' model. In this case to quantify the error in the system, the root mean square error (RMSE) of the water levels at every grid point and during the last 24 hours of simulation, has been computed. Moreover, the estimated standard deviation (STDEV) of the error in the water level at every grid point, has been obtained from the error covariance matrix calculated by the filter. The values of the STDEV represent the mean values during the last 24 hours simulation. Equations 4.3 and 4.4 indicate how the RMSE and the STDEV are calculated at point i. For this specific case, values of t_1 and t_2 are respectively 96 and 192 in time step units, this corresponds to the first and the last time steps of the second day of simulation.

$$RMSE_i = \sqrt{\frac{1}{(t_2 - t_1)} \sum_{t=t_1}^{t_2} \left(\zeta_{i,true}^t - \zeta_{i,wrong}^t\right)^2} \tag{4.3}$$

$$STDEV_i = \frac{1}{(t_2 - t_1)} \sum_{t=t_1}^{t_2} \sigma_{\xi_i} \tag{4.4}$$

Figure 4.2 shows the spatial distribution of the average RMSE of the wrong model and the average STDEV calculated with both filters.

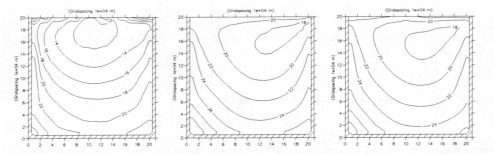

Figure 4.2: Left: average RMSE of the wrong model. Centre: average STDEV obtained from the RRSQRT using 40 modes and Right: average STDEV obtained from the EnKF using 100 ensemble member. RMSE and STDEV in cm.(without measurement update)

To have an idea of the global performance of the filters, it is possible to calculate the mean RMSE and STDEV throughout the entire domain. The mean values are 17.25, 22.1 and 21.6 cm., for the RMSE of the wrong model, the STDEV from the RRSQRT filter and the STDEV from the EnKF, respectively. Both filters provide similar results. They systematically overestimate the error but in general they can provide a good estimation of the error distribution as it can be observed in figure 4.2.

4.2.1 Perfectly known statistics

Though it is usual in a real data assimilation application that the error statistics are unknown or in the best case partly known, it is very important to test the filter under the hypothetical conditions of perfectly known error statistics. Under these conditions, the filter has to converge, reduce the system error and estimate correctly its own performance by means of the error covariance matrix. When all the statistical parameters are given, the only parameter that influences the filter convergence is the number of modes or number of ensemble members, M. It is considered that the filter has converged when increasing M the filter results do not present significant variations. In order to measure filter performance, the root mean square error (RMSE) between the 'true' model and the model updated by the filter is used. The RMSE is calculated, as the root mean square error of the water levels at every grid point during the last 24 hours of simulation, following a similar expression as equation 4.3

M	RMSE RRSQRT	STDEV RRSQRT	RMSE EnKF	STDEV EnKF
10	7.45	5.33	32.96	4.18
20	6.54	6.22	10.54	5.50
30	6.61	6.65	8.40	6.09
40	6.53	6.84	8.06	6.30
50	6.59	7.00	7.99	6.36
60	6.55	7.09	7.60	6.52
70	6.57	7.17	6.92	6.55
100	6.59	7.38	7.09	6.89
200	6.64	7.74	6.62	7.10

Table 4.1: Test 1. Filters performance with respect to the parameter M for the case of perfectly known error statistics. RMSE and STDEV in cm.

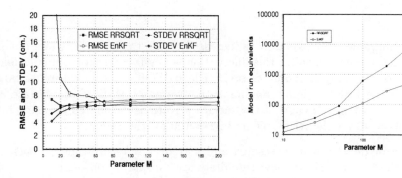

Figure 4.3: Left: RMSE and STDEV versus the parameter M. Rigth: Deterministic model run equivalents versus the parameter M for the RRSQRT filter and the EnKF

Table 4.1 and figure 4.3 show the values of the RMSE and the STDEV for both the RRSQRT and the EnKF as a function of M. The RMSE and the STDEV have the same value for values about 40 in the case of the RRSQRT and over 100 for the EnKF. The RRSQRT converges for a smaller value of M and to a slightly smaller value of the RMSE than the EnKF. After convergence both filters fluctuate

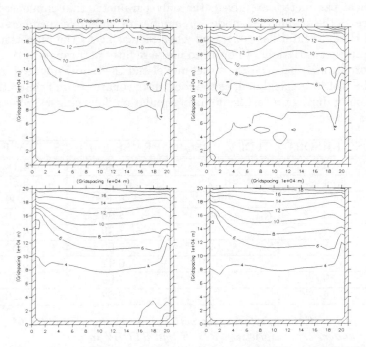

Figure 4.4: Test1. Left: from top to bottom average RMSE and STDEV calculated by the RRSQRT with M = 40. Right: from top to bottom average RMSE and STDEV calculated by the EnKF with M = 100. RMSE and STDEV in cm.

around the minimum value when M is increased, while the STDEV grows but not very significantly. The reason for this fluctuation is possibly due to accumulation of numerical errors (mainly rounding errors). The RRSQRT always estimates larger STDEV than the EnKF. Figure 4.4 shows the average RMSE of the surface elevations and the average STDEV of the surface elevations calculated by using both filters with $M = 40$ for the RRSQRT and $M = 100$ for the EnKF.

Both filters provide very similar results, and moreover the similitude between the RMSE and the STDEV shows how the filters are estimating quite accurately the error remaining in the system.

Regarding the computer requirements, both filters require M model integrations for the propagation of the error covariance matrix. Despite the faster convergence of the RRSQRT than the EnKF (i.e. need a smaller M), the computer time for the RRSQRT is very much dependent on the calculation of the eigendecomposition. Due to this factor, the computer time for a RRSQRT application grows in an exponentially manner, while for the EnKF the growth is nearly linear. Figure 4.3 presents computer time, expressed as equivalent deterministic model runs, as a function of M for the present test and for both the RRSQRT and the EnKF. The calculation have been carried out in a HP 9000/735/99.

In figures 4.5 and 4.6 are shown the time series of water levels at two measurement positions and two verification positions of the true model, the wrong model and obtained by both filters using $M = 40$ in the RRSQRT filter and $M = 100$ in the EnKF. It can be observed that the filters are able to recover very accurately the water levels at measurement points as well as at other points of the model.

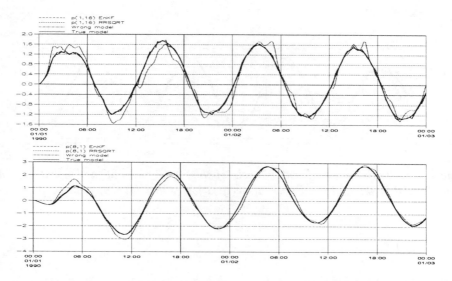

Figure 4.5: Time series of water levels of the 'true', 'wrong' and updated models from the RRSQRT and the EnKF at measurements points (1,16) and (8,1).

It is easy to observe that both filters not only reduce significantly the RMSE in most parts of the area but also they can predict very accurately the model errors. Larger errors remain in the area close to the open boundary because the measurements have less influence there despite the fact that the boundary is corrected at every time step using the estimated values of the boundary errors obtained in the augmented part of the state vector (see section 3.1.2). Figure 4.7 shows the time series, at one specific point of the boundary, of the error estimated by both filters and the noise added to the boundary to generate the wrong model (the true noise). It is observed that the estimation has a phase error, i.e. the filter estimates the error a bit late, though in general the main trend of the error is captured by the filter.

The case that has been tested above was selected using coloured noise defined by a first order autoregressive process with lag-one autocorrelation coefficient equal to 0.9. The performance of the filter is different when the 'wrong' model is generated using smaller autocorrelation coefficients. Figure 4.8 shows the filter performance when the noise has been generated using processes with $\sigma = 0.218$ and $\alpha = 0.5$, and $\sigma = 0.272$ and $\alpha = 0.0$ respectively. In both cases the wrong model has the

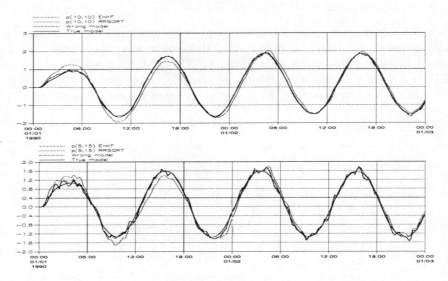

Figure 4.6: Time series of water levels of the 'true', 'wrong' and updated models from the RRSQRT and the EnKF at verification points (10,10) and (5,15)

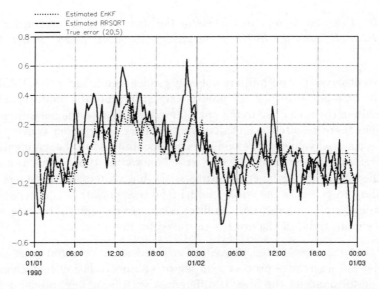

Figure 4.7: Time series of the true error and estimated from the RRSQRT and the EnKF at one boundary point

same value of the spatially averaged RMSE as in the main test considered at the beginning of this section. The comparison between the results from figure 4.8 with the result from figure 4.3 shows that when the noise is generated from a process without significant temporal correlation, both filters converge for a larger value of M and to a larger value of the RMSE.

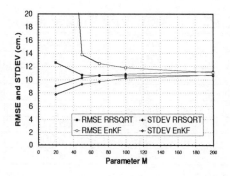 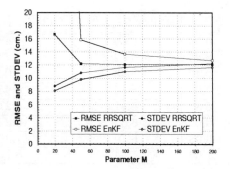

Figure 4.8: Test 1. RMSE and STDEV versus the parameter M using left: $\sigma = 0.218$ and $\alpha = 0.5$; right: $\sigma = 0.272$ and $\alpha = 0.0$

4.2.2   Sensitivity and robustness

The case, which has been studied in the previous section, where the noise statistics are perfectly known, it is extremely rare to happen in a real application. Usually, the knowledge of the statistics of the error is very limited. Therefore it is very important to know how the filter performs when the error statistics are incorrectly defined. Beside the parameter M, the most influential parameters for the filter, which are going to be investigated here are:

1. Temporal correlation

2. Spatial correlation

3. System noise and measurement noise variances

Moreover, the filters are going to be tested when a systematic error is introduced, for example an error in the phase or in the amplitude of the boundary conditions. The parameter M has been chosen differently for each filter based on convergence considerations. In the case of the RRSQRT $M = 40$ is going to be used, while for the EnKF M is chosen equal to 100. The other parameters are kept as in section 4.2.1.

Temporal correlation

The wrong model was generated by adding noise to the boundary conditions that is correlated in time using equation 4.1 and with a value of the lag-one autocorrelation coefficient α equal to 0.9. In order to test the sensitivity of the filter with respect to the temporal correlation, the same test has been run using different values of the parameter α. Test results are shown in table 4.2 based on the parameters RMSE and STDEV as introduced in the previous section.

α	RMSE RRSQRT	STDEV RRSQRT	RMSE EnKF	STDEV EnKF
0.00	6.54	4.48	7.52	4.70
0.20	6.52	4.73	7.11	4.85
0.40	6.43	5.03	6.83	5.10
0.60	6.34	5.45	6.70	5.50
0.80	6.41	6.21	6.82	6.21
0.90	6.53	6.84	7.09	6.84
0.95	6.70	7.39	7.38	7.36
1.00	7.54	8.72	8.30	8.25

Table 4.2: Test 1. Filters performance with respect to the temporal correlation α. RMSE and STDEV in cm.

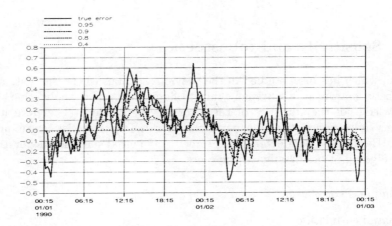

Figure 4.9: Time series of error at one boundary point calculated with different values of α

The results show that the global performance of both filters in terms of RMSE, is not strongly influenced by the definition of α, though the EnKF is slightly more sensitive to changes in this parameter. Both filters are able to correct the model output as observed by the similar values of the RMSE. The error estimation represented by the STDEV varies more significantly when α is modified, therefore the error remaining in the model is not well estimated. In this case both filters provide almost the same values of the STDEV for each value of M. The estimation of the error at the boundary is more sensitive to a misspecification of α. Figure 4.9 shows the boundary error estimation at one point of the boundary for different values of α. For $\alpha < 0.6$ the filter is not able to recover the boundary noise. The filter estimates the boundary noise more correctly when the value of α is close to the real value, in this case 0.9.

Spatial correlation

Errors in the wrong model have been generated using a Gaussian distribution with zero mean and covariance matrix Q_w given by the equation 4.2. Sensitivity of the filter performance when the correlation coefficient ρ is incorrectly defined is tested here. All the statistical parameters but the spatial correlation are considered known. The filters performance using different values of the spatial correlation ρ are shown in table 4.3

ρ	RMSE RRSQRT	STDEV RRSQRT	RMSE EnKF	STDEV EnKF
0.00	7.97	3.75	8.05	4.99
0.20	7.62	3.83	7.81	5.31
0.40	7.10	4.52	7.54	5.70
0.60	6.69	5.73	7.28	6.20
0.80	6.51	6.62	7.09	6.73
0.90	6.53	6.84	7.09	6.84
0.98	6.82	6.47	7.38	6.19

Table 4.3: Test 1. Filters performance with respect to the spatial correlation ρ. RMSE and STDEV in cm.

Results in table 4.3 show that the global model performance is not very sensitive to changes in the spatial correlation. Moreover, the results get worse as ρ gets closer to zero, at the same time the STDEV decreases, with the consequent deterioration in the estimation of the model errors. The RRSQRT is more sensitive than the EnKF in the STDEV estimation, than the EnKF, but at the same time it has a smaller RMSE than the EnKF for all the range of values of ρ.

System noise and measurement noise variances

The influence of these two parameters on the filters performance, cannot be considered independently, because the ratio between the system noise and the measurement noise has an extreme importance for filter stability. Table 4.4 shows the filter performances using different values of the system noise standard deviation. The STDEV varies very rapidly with σ. When the STDEV reaches large values, the RRSQRT filter become unstable, i.e. cannot produce any result, while the EnKF produce unstable results, i.e. the results present significant oscillations.

σ	σ/σ_m	RRSQRT		EnKF	
		RMSE	STDEV	RMSE	STDEV
0.01	0.2	7.83	1.69	7.73	1.38
0.05	1	6.59	4.05	6.91	3.99
0.10	2	6.53	6.84	7.09	6.84
0.20	4	7.02	12.40	8.03	12.25
0.40	8	9.75	32.15	10.85	22.46
0.50	10	-	-	12.52	27.32

Table 4.4: Test 1. Filters performance with respect to the magnitude of the standard deviation σ of the model error and the ratio σ/σ_m. RMSE and STDEV in cm.

Using different values of the standard deviation of the measurement noise, the performance of the filters do not suffer important variations while the ratio σ/σ_m is not very large. Also in this case, the EnKF can deal with large values of the ratio, though producing unstable solutions, while the RRSQRT cannot even provide a solution

σ_m	σ/σ_m	RRSQRT		EnKF	
		RMSE	STDEV	RMSE	STDEV
0.005	20	-	-	20.98	5.25
0.010	10	-	-	11.87	5.60
0.015	6.7	8.59	5.92	9.47	5.81
0.050	2	6.53	6.84	7.09	6.84
0.100	1	6.58	7.95	6.95	7.96

Table 4.5: Test 1. Filters performance with respect to the magnitude of the standard deviation σ_m of the measurement error and the ratio σ/σ_m. RMSE and STDEV in cm.

The value of the ratio σ/σ_m, (Cañizares [8]) indicates if the updated state is based more on the model or on the measurements. The filter is unstable when the ratio becomes large (values around 7-8 for the RRSQRT in our simple case), i.e. when the filter output is strongly based on the measurements.

Systematic errors at the boundary

In most of the real applications the forcing contains systematic errors that are not included in the filter error statistics. The typical systematic errors are phase errors and errors in the amplitude. In these tests, the filter is provided with the noise statistics, which do not include the systematic error. Table 4.6 shows the filters results when phase errors of 1 and 3 hours were included in the boundary. Filter corrections are extremely significant, with a very important reduction in the RMSE.

phase	RMSE Wrong	RRSQRT		EnKF	
		RMSE	STDEV	RMSE	STDEV
+ 1h	90.72	9.28	6.87	9.74	6.86
+ 3h	186.66	18.3	6.94	19.01	6.90

Table 4.6: Filters performance with respect to phase errors at the boundary. RMSE and STDEV in cm.

At the same time, both filters are able to recover most of the error at the boundary. Figure 4.10 shows the time series of water levels at the verification point (5,15) from the 'true', the 'wrong' and the updated model with the filter. Figure 4.11 shows the error at one point of the boundary when a phase error of 3 hours has been included, together with the filter estimation.

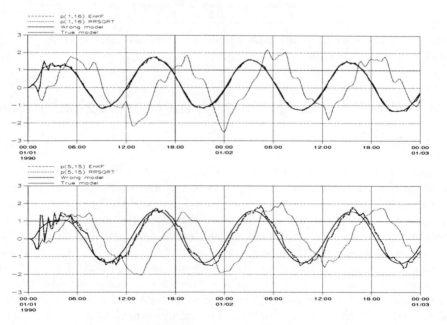

Figure 4.10: Time series of water levels of the 'true', 'wrong' and updated models from the RRSQRT and the EnKF at two different locations for the test case of phase error 3 hours. From top to bottom: measurements at (1,16) and verification point (5,15)

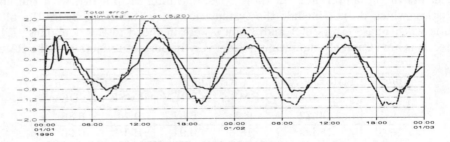

Figure 4.11: Time series of error at one boundary point, phase error 3 hours

Table 4.7 presents the results from a test where the range of the water level at the boundary has been increased from 2 meters to 3. The two filters provide very good results, not only by a global error reduction in the model but also by a good estimation of the amplitude error at the boundary.

Amplitude	RMSE Wrong	RRSQRT		EnKF	
		RMSE	STDEV	RMSE	STDEV
3 m.	62.84	8.81	6.86	9.46	6.81

Table 4.7: Filters performance with respect to amplitude errors at the boundary. RMSE and STDEV in cm.

4.3   TEST 2: Error in the meteorological forcing terms

The true model introduced in section 4.1 has been also used for this test. This model has been disturbed by adding errors to the meteorological forcing terms in the momentum equations. These errors have been randomly generated, as in the boundary noise case, from a first order autorregresive process with a lag-one auto-correlation coefficient $\alpha = 0.97$ and residual errors δ_k generated from a Gaussian distribution with zero mean and covariance matrix Q_w given by the expression 4.2 using an exponential distance correlation with correlation coefficient $\rho = 0.98$ and l the distance between points i and j. These parameters are the same as those used by Heemink [33] who determined them by a combination of physical intuition and trial and error for the filter calibration. The standard deviation σ is constant and equal to $0.0005\ m^2/s^2$. As in Test 1, the state obtained from this simulation is called the 'wrong state', and is used as the model forecast. The boundary used for the wrong model is the same as that used for the true model, therefore it is assumed that the boundary is perfectly known. In section 3.5.4 the scaling problem was already introduced, therefore in this case the eigenvalue decomposition is calculated for the normalised error covariance matrix.

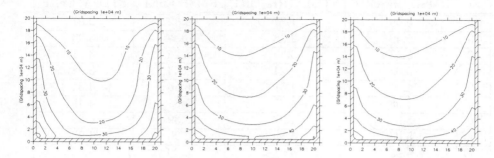

Figure 4.12: Test 2. Left: average RMSE of the wrong model. Centre: average STDEV obtained from the RRSQRT using 70 modes and Right: average STDEV obtained from the EnKF using 200 ensemble member. RMSE and STDEV in cm. (without measurement update)

As in the previous section, the two filters are initially compared with respect to the error covariance estimation when no measurements are assimilated. The filter is provided with the same noise statistics used to generate the 'wrong' model. The same parameters (RMSE and STDEV) described in equations 4.3 and 4.4 and applied in the boundary noise case are used here. The value of the parameter M used for this test is equal to 70 for the RRSQRT and 200 for the EnKF. Figure 4.12 shows the spatial distribution of the RMSE of the wrong model and the average STDEV calculated with both filters. The spatially averaged values are for this test 17.91, 20.98, and 20.64 *cm.*, for the RMSE of the wrong model and the STDEV of the RRSQRT and the EnKF respectively. It can be observed that using the correct statistics, both filters are able to forecast quite accurately the error in the model, though they overestimate the error. Despite they produce very similar results, the RRSQRT overestimate slightly more than the EnKF.

4.3.1 Perfectly known statistics

Under the assumption of perfectly known statistics the filters performance have been tested with respect to the parameter M, i.e. the number of modes in the RRSQRT and the number of ensemble members for the EnKF.

Table 4.8 presents the values of the spatially averaged RMSE and STDEV for both filters and for different values of M. The values of the RMSE as well as the STDEV converge very fast and to a smaller value for the RRSQRT than in the case of the EnKF. Filters behaviour are similar to the boundary case, in the sense that after convergence the RMSE fluctuate around the minimum value while the STDEV slightly grows. Figure 4.13 shows the aforementioned results in graphical form, where it is easy to observe the difference between the two filters with respect to their convergence.

M	RMSE RRSQRT	STDEV RRSQRT	RMSE EnKF	STDEV EnKF
10	6.09	1.77	9.82	2.07
20	5.40	2.38	10.24	3.03
30	5.15	3.35	8.26	3.61
40	5.08	4.24	6.03	3.98
50	4.91	4.51	6.46	4.31
60	4.98	4.75	5.88	4.45
70	5.03	4.97	5.39	4.35
100	5.03	5.19	6.90	4.65
200	4.93	5.51	5.00	5.02

Table 4.8: Test 2. Filters performance with respect to the parameter M. RMSE and STDEV in cm.

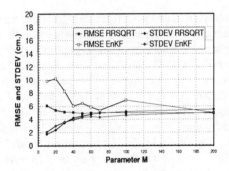

Figure 4.13: RMSE and STDEV versus the parameter M for the default case in Test 2

With regards to computer requirements the same conclusion as in Test 1 can be made, however it should be pointed out that since more noise points are used in this case, the computation of the eigenvalue decomposition in the RRSQRT filter is a little bit more expensive than in the boundary test. Therefore in this case, for the same value of M, the RRSQRT becomes more computationally demanding than the EnKF.

Despite that both filters, and especially the RRSQRT, produce significant corrections in the model for small values of M (in terms of small RMSE), the system error estimation in those cases is very poor because the STDEV is very small. Only after the STDEV has reached similar values to those of the RMSE, is the error estimated quite accurately. This can be observed in figure 4.14 where the spatial distribution of the RMSE and the STDEV for the RRSQRT using $M = 70$ and for the EnKF using $M = 200$ are presented. It can be observed that both filters accurately estimated

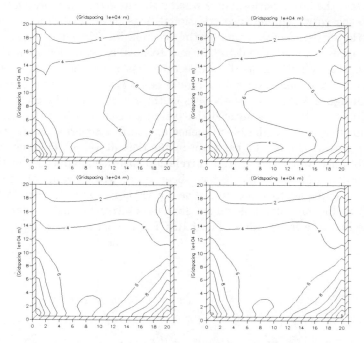

Figure 4.14: Test 2. Left: from top to bottom average RMSE and STDEV calculated by the RRSQRT with M = 70. Right: from top to bottom average RMSE and STDEV calculated by the EnKF with M = 200. RMSE and STDEV in cm.

the error in the system, i.e. the STDEV (estimation of the error in the system) is very similar to the RMSE (error remaining in the system).

For the RRSQRT application it was necessary to normalise the error covariance matrix, because the error associated with the DMO error variables (this type of error was introduced in the last part of the previous chapter) has a significantly smaller magnitude than other variables of the state vector. The modes associated with the DMO error fields are the most important in the error covariance matrix. Therefore, even in the case where only 10 modes have been used, the estimation of those errors was very similar to the one obtained with 70 modes. This indicates that the first few modes of the error covariance matrix are very important for the DMO error estimation. The same result is observed in the EnKF where the error estimated using only 20 ensemble members is reasonably similar to the one obtained with 200. The same behaviour was observed in Test 1, i.e. the error at the boundary were estimated using small values of M. Figure 4.15 shows the time series at location (10,10) of the error used to generate the 'wrong' model and its estimation by the RRSQRT using 70 modes, and by the EnKF using 200 modes. Both errors associated with the momentum in the x- and in the y-directions are presented.

In order to test the filters performance when different error statistics have been used to generate the wrong model, two additional tests have been done. The effect of generating the statistics with a smaller time correlation, the lag-one autocorrelation coefficient, α has been set equal to 0.5, and to have a wrong model with a similar RMSE than the default value the value of σ has been chosen equal to 0.00092 m^2/s^2. Using these values the wrong model has a RMSE equal to 17.96 cm. The left graph in figure 4.16 shows the values of the RMSE and STDEV as a function of M, for this case. It can be noticed that using a value of α half that the one used in the default test has a negligible influence on the filters performance. Both filters converge to a value slightly higher than in the default value. Different conclusion has been obtained when the spatial correlation has been reduced. In this case a value of $\rho = 0.5$ and $\sigma = 0.00106$ have been used, with a value of the RMSE of the wrong model equal to 17.83 cm. The EnKF is not able to correct the model significantly and is quite unstable for small values of M, while using a large ensemble it converges but to a large value. The RRSQRT is unstable and it provides worse

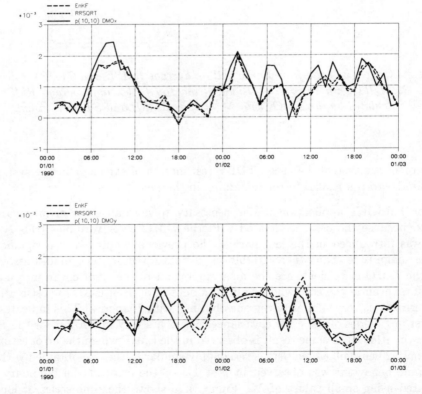

Figure 4.15: Time series at point (10,10) of the true and estimated error in the momentum equations (x and y directions)

results than the wrong model. The small spatial correlation generates smaller values of the standard deviation at points located among noise points, therefore corrections are very irregular and produce model instabilities. The test results are shown in the right graph of figure 4.16, where the vertical scale has been modified with respect to the other cases.

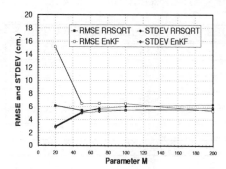 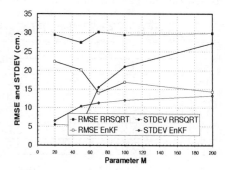

Figure 4.16: Test 2. RMSE and STDEV versus the parameter M using different statistical parameters to generate the wrong model. Left: smaller time correlation, right: smaller spatial correlation

4.3.2   Sensitivity and robustness

The filters performance when the noise statistics are not perfectly defined is studied in this section. A similar procedure to Test 1 is followed here. All the parameter but one, from which its sensitivity to the filters performance is going to be studied, are going to be maintained as in the default test. The parameters to be studied here are:

1. Temporal correlation

2. Spatial correlation

3. System noise and measurement noise variances

Moreover, the filters are going to be tested when a systematic error is introduced. In this case no meteorological forcing is provided to the model.

The parameter M have been chosen different for each filter based on convergence considerations as it was done for Test 1. In this case, the RRSQRT uses $M = 40$ and for the EnKF M is chosen equal to 100. The other parameters are kept as in the case of perfectly known statistics.

Temporal correlation

The wrong model was generated by adding noise at the DMO terms using a temporal correlation that follows equation 4.1, with a value of the lag-one autocorrelation coefficient α equal to 0.97. The sensitivity of the filter performance with respect to the temporal correlation has been tested here. The test consists of using the same parameters as in the default test but modifying the parameter α. The test results based on the parameters RMSE and STDEV are presented in table 4.9.

α	RMSE RRSQRT	STDEV RRSQRT	RMSE EnKF	STDEV EnKF
0.00	7.25	1.78	7.33	1.65
0.20	6.36	1.94	6.51	1.83
0.40	5.71	2.17	5.83	2.07
0.60	5.32	2.49	5.39	2.41
0.80	5.17	3.03	5.18	2.97
0.90	5.20	3.57	5.19	3.56
0.97	5.03	4.97	5.00	5.02

Table 4.9: Test 2. Filters performance with respect to the temporal correlation α. RMSE and STDEV in cm.

The filters behaviour is similar to Test 1. The global performance of both filter in terms of RMSE is not strongly influenced by the value of α, though the results slightly deteriorates for very small values of α. The error estimation, represented by the STDEV is poor if the parameter α is smaller than 0.90. Moreover, the estimation of the errors in the DMO terms is also very bad for the same cases. The sensitivity of both filters to the parameter α in the case of errors in the meteorological forcing is the same as in the case of errors at the boundaries. It can be concluded, that both filters present the same behaviour with respect to variations in the temporal correlation.

Spatial correlation

The filters sensitivity when the correlation coefficient ρ is incorrectly defined is tested here. All the statistical parameters but the spatial correlation are considered known. The filters performance using different values of the spatial correlation ρ are shown in table 4.10.

In contrast to Test 1 both filters are more sensitive to the definition of the spatial correlation. This is due to the spatially distributed nature of the errors in the meteorological forcing. In this specific test, the EnKF seems to be more sensitive with respect to state corrections (larger RMSE for values between 0.6 and 0.9) and to error estimation (it consistently overestimate the errors, i.e. provides large STDEV). Both filters present a sudden increase of the STDEV for values of ρ between 0.8 and 0.9, and a decrease for the default value of $\rho = 0.98$.

ρ	RMSE RRSQRT	STDEV RRSQRT	RMSE EnKF	STDEV EnKF
0.00	10.90	3.24	10.47	6.23
0.20	10.86	3.25	10.46	6.23
0.40	10.12	3.60	10.31	6.35
0.60	7.40	4.90	9.52	6.87
0.80	5.76	6.92	7.26	7.89
0.90	4.97	7.43	5.62	7.81
0.98	5.03	4.97	5.00	5.02

Table 4.10: Test 2. Filters performance with respect to the spatial correlation ρ. RMSE and STDEV in cm.

System noise and measurement noise variances

The filters sensitivity to an incorrect definition of the standard deviations of the DMO and measurement noise, is tested here. Table 4.11 presents the test results using different values of the standard deviation of the noise in DMO.

σ	RMSE RRSQRT	STDEV RRSQRT	RMSE EnKF	STDEV EnKF
0.000025	7.79	1.31	7.84	1.10
0.000100	5.36	2.51	5.42	2.48
0.000250	5.03	4.97	5.00	5.02
0.000500	4.54	9.10	5.24	9.19
0.001000	6.22	17.11	7.36	17.21
0.002000	15.14	31.10	12.50	32.54

Table 4.11: Test 2. Filters performance with respect to the magnitude of the standard deviation σ of the model error. RMSE and STDEV in cm.

Both filters showed a very similar behaviour. They produce good model corrections excpt for cases of extremes values of σ, i.e. for very small or very large values of σ. Moreover, they are able to estimate the model error only in the cases where σ is very close to the real value. Other cases produce either overestimation or underestimation of the error.

Regarding filters sensitivity to measurement errors, results are presented in table 4.12.

As in the previous case both filters have a very similar behaviour. For very small values of the standard deviation of the measurement noise, both filters are not able to correct the system. When a large standard deviation is used the filters can correct the model but they overestimate the errors.

σ_m	RMSE RRSQRT	STDEV RRSQRT	RMSE EnKF	STDEV EnKF
0.005	15.08	3.87	18.59	3.99
0.010	8.86	4.13	9.33	4.27
0.050	5.03	4.97	5.00	5.02
0.100	5.25	5.70	5.34	5.80
0.200	5.76	7.23	5.80	7.33

Table 4.12: Test 2. Filters performance with respect to the magnitude of the standard deviation σ_m of the measurement error. RMSE and STDEV in cm.

Systematic errors: Zero meteorological forcing

It was already commented upon that in most of the real applications, the forcing contains systematic errors. A new test has been set-up in order to test both filters under the existence of systematic errors. The test consists of neglecting the meteorological forcing terms. In this case no random errors have been introduced. The same measurements obtained from the true model defined at the beginning of this section have been used. Table 4.13 presents the test results for both filters.

Amplitude	RMSE Wrong	RRSQRT		EnKF	
		RMSE	STDEV	RMSE	STDEV
No wind	19.26	5.76	7.45	5.91	5.17

Table 4.13: Filters performance with respect to amplitude errors at the boundary

It can be concluded that in this case, both filters significantly correct the model. The EnKF estimates more accurately the error remaining in the system than the RRSQRT, though both are quite accurately. It has to be observed here that the model is driven mainly by the boundary conditions and that the error produced by neglecting the meteorological forcing terms is not as critical as in the test case of systematic errors in the boundary condition.

4.4   Some remarks about the mass conservation problem

Under the usual assumption of model errors due to uncertainties in the forcing terms, the filter has none or very few degrees of freedom to generate changes in the system state. When only noise in the momentum equation is considered, the mass (or volume) in the system is completely conserved. The filter reallocates water in order to produce the best agreement between model results and measurements. A direct consequence of that action is that in areas slightly influenced by the measurements the result deteriorates. Water is taken from or accumulated in those areas to correct the areas of influence of the measurements.

To illustrate this problem a similar model to the one used in previous tests but closing the open boundary is used here. The model is forced by the same cyclonic wind used in previous sections. The meteorological forcing is perturbed by adding the same errors in the momentum equation as in Test 2, i.e. with parameters $\alpha = 0.97$, $\rho = 0.98$ and $\sigma = 0.00025 \ m^2/s^2$. Simultaneously, the initial conditions are strongly modified. The true model has a static initial state (velocity equal zero everywhere) and a constant surface elevation of $0.0 \ m$. The wrong model is forced using a meteorological forcing perturbed with the aforementioned errors and its initial state is also static but with the surface elevation at level -0.25 m.

The RRSQRT filter has been used to correct the model, using the same measurement positions defined in Test 2, and using a value of M equal to 70. Assuming only noise in the momentum equation, the filter solution consists of displacing the water to or from the measurement positions in order to produce very good agreement on water levels at the measurement positions and also at neighbouring points. In figure 4.17 is is shown the RMSE of the wrong model, which has a spatially averaged value of 27.88 cm. The RMSE obtained from the previously defined RRSQRT application is also shown in figure 4.17 together with the STDEV. The average values of these parameters are respectively 26.65 and 4.94 cm. It can be concluded from these results and also from figure 4.18 that the filter provides good estimates of the water levels at the measurement positions but worse estimates at other locations. Consequently, the overall results cannot be considered as acceptable.

Under the assumption of errors in the water levels, the filter has the necessary degrees of freedom to modify the overall mass (or volume). In this case, the water does not have to be reallocated but added or taken by the filter whenever it is needed. The spatially averaged RMSE in this case is 7.18 and the STDEV 8.13, therefore the filter is able to estimate the error quite accurately. It is easy to observe the similitude between the RMSE and the STDEV in figure 4.17 in this case. Moreover, in figure 4.18 it can be observed in both the measurement position and in the verification position, that the filter is able to produce very good corrections when errors in the water levels have been considered. On the other hand, it is also shown that in the case when only errors in the momentum equation have been considered, the result at the verification points is even worse than the one obtained from the wrong model.

To have a measure of the overall correction, the true model has a constant volume of $14300 \times 10^8 m^3$ and the wrong model $14205 \times 10^8 m^3$. Using only errors in the momentum equation the final result has the same volume as the wrong model while using errors in the water levels, the final volume obtained is $14286 \times 10^8 m^3$.

4.5 Conclusions

In this chapter the implementation in a two-dimensional hydrodynamic model of two data assimilation schemes based on the Kalman filter algorithm have been tested. The model noise process is related to model errors through the open boundary

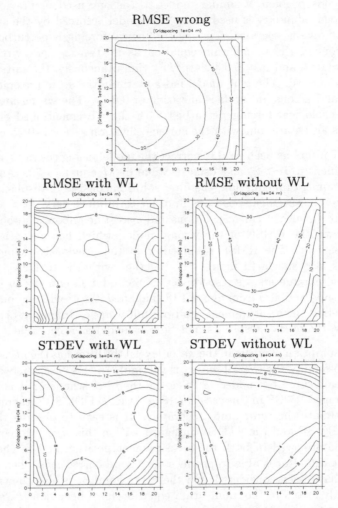

Figure 4.17: Top: average RMSE of the wrong model. Left:from top to bottom average RMSE and STDEV obtained from the filter considering noise in the water levels. Right: from top to bottom average RMSE and STDEV obtained from the filter using only noise in the momemtum equation.

conditions and the meteorological forcing. An augmented state vector formulation is adopted in order to implement time coloured model noise. The Kalman filter provides simultaneously corrections of the system state and the forcing terms during the assimilation procedure.

A twin experiment based on a hypothetical bay region has been used to test the performance of the models. Some conclusions obtained from the tests are:

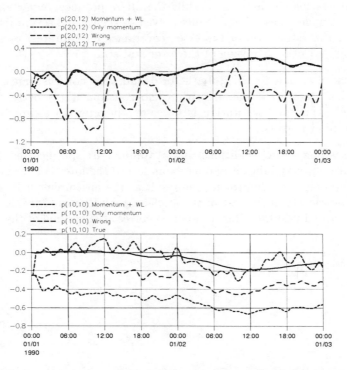

Figure 4.18: Time series at measurement point (20,12) (top) and verification point (10,10) (bottom) of the true and wrong models and the filter results using only noise in the momentum equation and using also noise in the water levels.

- The most important parameter is the rank reduction (number of leading eigenvalues) for the RRSQRT filter and number of ensemble members for the EnKF. The reduced rank representation of the RRSQRT is more efficient than the ensemble representation in the EnKF, in the sense that the same result could be obtained using a smaller number of leading eigenvalues than ensemble members. However, the eigenvalue decomposition used in the RRSQRT filter is very time consuming increasing the computational load of this filter. Therefore, both filters provide similar results for an equivalent computational load.

- When the correct noise statistics are provided to the filters, the estimated error covariance is very similar to the observed RMSE. This can be achieved using a very small number of leading eigenvalues or ensemble members compared with the dimension of the state vector.

- In real applications the error statistics are unknown or only partly known, therefore the robustness of the filters with respect to misspecifications of the error statistics has been tested. In general, both filters have shown a significant robustness for all the tests. While the filters produce very good model corrections under these circumstances (even in the case of severely biased model errors) the estimation of the error covariance deteriorates, presenting more sensitivity to the definition of the error statistics.

- The use of an augmented state vector defining the noise as time coloured noise provides a more efficient algorithm and consequently a better filter performance. This implementation is necessary in order to estimate and correct biased model errors.

- When wrong initial conditions (initial volume) are specified for closed models (or with not very influential open boundaries) the filter is not able to produce global corrections. Improving the result at the measurement locations causes it to deteriorate in the rest of the model. More degrees of freedom have to be incorporated into the filter in order to correct the volume of the system.

Chapter 5

Application on a 2D hydrodynamic model. Test case

Despite the increasing development and use of three-dimensional hydrodynamic models, two-dimensional models are still one of the main constituents in many engineering applications and specially in storm surge models. Bode and Hardy [4] review the topic of storm surge models, with a description of the most important components of the model, as well as defining the state of the art within this topic. They consider 2D models as the backbone of state of the art operational storm surge prediction for which coastal water levels are of major concern. The result of these models do not always match the observations, due to all the simplifications introduced. Real-time assimilation of data into storm surge models has been used during the last years to improve the model results. Applications of the Steady state Kalman filter (Heemink [31]) on linear 2D shallow water models for storm surge forecasting can be found in Bolding [5] Heemink [32], De Vries [66], Vested et al. [65] and Heemink et al. [34]. The last reference also includes an application of the RRSQRT filter on a twin test for storm surge forecasting in the North Sea. The RRSQRT filter permits the use of a set of observations that are variable in time and space, which is very convenient for data assimilation in storm surge modelling. The RRSQRT filter has been tested using a real case, the storm occurred on the North Sea during February 1993, and preliminary results can be found in Cañizares et al. [10].

5.1 The North Sea model test case

5.1.1 The model set-up

The North Sea is one of the most studied marine areas in the world. This is a very important feature when testing the methods, because of the availability of information. In order to obtain better deterministic model results, it will be necessary to use a fine grid resolution model. The high computational cost associated with a data assimilation application makes necessary the use of a coarser grid during the research period. In this way it will be possible to carry out several tests. At the same time the grid size has to be fine enough to reproduce the main flow patterns in the model.

Under previous assumptions a bathymetry with a grid size of 9 nautical miles (16670 m.) in both directions, has been selected. This bathymetry has been extracted from the DYNOCS model (dynamics of connecting seas) [36] and modified by closing the area of the Danish Belts. The bathymetry is shown in figure 5.1.

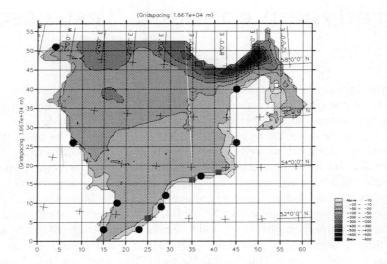

Figure 5.1: North Sea model bathymetry (Depth in meters) and available water level stations represented with circles and squares.

At the two open boundaries, the water levels at each grid point and for each time step are specified by the sum of the astronomical tidal level and the atmospheric pressure forcing level component (ζ_p). The astronomical tidal level for the Northern boundary, is calculated from the ten following tidal constituents: $M_2, S_2, N_2, K_2, \upsilon_2, K_1, O_1, P_1, M_4$ and MS_4. In the case of the southern boundary the predicted tide was calculated using the admiralty tables at the closest stations (Dungeness and Wissant) where the open boundary is located. The influence of atmospheric pressure in the open boundaries is calculated using the following semi-empirical expression:

$$\zeta_p = (P_n - P) \times 0.0001 \tag{5.1}$$

where:

ζ_p Change in the water level caused by the atmospheric pressure (m)
P Atmospheric pressure (Nm^{-2})
P_n Neutral pressure $(101{,}300\ Nm^{-2})$

Further model simplifications are: the earth's curvature and the tide generating forces are neglected and the density is assumed constant in time and in space.

The flow resistance is define with a space variable Manning number as a function of the water depth. The values are shown in table 5.1

Depth	Manning coef. $m^{1/3}/s$
> 150	4
150 - 100	18
< 100	32

Table 5.1: Manning coeffient used as a function of the depth

Meteorological data have been calculated by the HIRLAM model that runs operationally at DMI (Danish Meteorological Institute). The wind velocity fields as well as pressure fields are obtained from HIRLAM every 6 hours, and they are interpolated to every model time step. An example of the meteorological input is shown in figure 5.2.

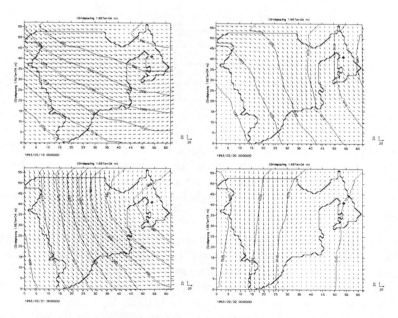

Figure 5.2: Pressure and Wind velocity fields, every 24 hours from 19/03/93 to 22/03/93

5.1.2   Deterministic model performance

The deterministic model performance has been first tested against the predicted tides calculated from the admiralty method at 13 different locations. These locations have been chosen because water level data from all of them are available for the period of study: from 18/02/93 00:00 to 23/02/93 00:00. The water level data as well as the meteorological data presented in previous sections have been obtained from the ECAWOM MAS2-CT94-0091 project.

Geographically the data is divided into three areas; the British coast, the Dutch coast and the eastern Danish coast. Table 5.2 presents the 13 stations and their co-ordinates in the model grid. Figure 5.1 shows the location of the 13 measurement stations in the model domain.

Station	x-direction	y-direction
Wick	4	51
North Shields	8	26
Lowestoft	18	10
Sheernes	15	3
Vlissingen	23	3
Hoek van Holland	25	6
IJmuiden	28	9
Den Helder	29	12
Harlingen	35	16
Huibertgat	37	17
Delfzijl	41	18
Esbjerg	45	26
Hantsholm	45	40

Table 5.2: Measurement stations and their co-ordinates in the model grid

First, a simulation without any meteorological forcing and using only the tidal variation at the two open boundaries was carried out. Simulation results can be observed in table 5.3, together with the predicted tides obtained from the admiralty method at the 13 stations. The simulated and predicted range of the tides during the five-day period for every station are also included.

Furthermore, table 5.3 presents the standard deviation of the difference between simulated and observed water levels for the 5-day period considered. Moreover, the relative error to tidal range at every station is also included. The model has in general a good agreement for all the stations with small error relative to the tidal range. The worst performance takes place in Hanstholm where the deterministic tide has a range larger than the predicted tide. Note that this station has a strong influence from the Northern Boundary and is very sensitive to small errors in phase as well as in the amplitude of the incoming waves. Figure 5.3 shows the time series of measured and modelled tidal water levels at six stations for the five-day period

Station	Stdev.error cm./% range	Predicted tide		Deterministic tide	
		Min (m.)	Max (m.)	Min (m.)	Max (m.)
Wick	0.24/8.4	-1.41	1.44	-1.54	1.39
North Shields	0.18/4.3	-2.21	2.13	-2.05	1.84
Lowestoft	0.15/7.1	-1.13	0.93	-0.85	0.67
Sheernes	0.35/8.0	-2.21	2.22	-2.03	2.06
Vlissingen	0.25/5.9	-2.00	2.24	-1.63	2.05
Hoek van Holland	0.10/5.3	-0.80	1.12	-0.79	1.29
IJmuiden	0.23/12.5	-0.79	1.03	-0.64	1.09
Den Helder	0.13/8.0	-0.84	0.74	-0.86	1.00
Harlingen	0.13/6.4	-0.93	1.03	-0.89	1.11
Huibertgat	0.25/10.3	-1.38	1.07	-0.95	1.07
Delfzijl	0.19/6.1	-1.67	1.46	-1.05	1.21
Esbjerg	0.10/5.9	-0.89	0.79	-0.69	0.95
Hantsholm	0.12/39.1	-0.16	0.15	-0.29	0.34

Table 5.3: Comparison between predicted and deterministic tide

considered above. It can be observed that the computed water levels for all the stations present a very good agreement with the predicted values.

A second simulation was carried out including the meteorological forcing. As in the tidal case, the deterministic model results have been compared with observations at 13 stations and the results are presented in table 5.4 and figure 5.4.

Station	Stdev.error cm./% range	Observed WL		Deterministic WL	
		Min (m.)	Max (m.)	Min (m.)	Max (m.)
Wick	0.28/8.8	-1.75	1.38	-1.70	1.28
North Shields	0.30/6.0	-2.47	2.55	-2.04	2.37
Lowestoft	0.28/7.8	-1.25	2.34	-0.84	2.24
Sheernes	0.63/10.6	-2.53	3.42	-1.61	3.08
Vlissingen	0.63/11.4	-2.08	3.43	-1.27	2.87
Hoek van Holland	0.36/10.1	-1.06	2.55	-0.72	2.15
IJmuiden	0.34/9.3	-1.09	2.55	-0.40	2.40
Den Helder	0.26/6.8	-1.11	2.66	-0.54	2.29
Harlingen	0.40/9.0	-1.20	3.30	-0.72	2.48
Huibertgat	0.37/9.6	-1.62	2.20	-0.73	2.30
Delfzijl	0.67/12.5	-2.07	3.30	-0.83	2.62
Esbjerg	0.39/11.6	-1.38	1.96	-0.79	2.64
Hantsholm	0.23/15.2	-0.52	1.01	-0.42	0.95

Table 5.4: Comparison between observed and deterministic wate levels

The deterministic model in this case presents as expected, larger errors than in the tidal case. Large errors with standard deviation of 0.63 cm. are observed in the most southern stations (Sheernes and Vlissingen) probably due to the lack of

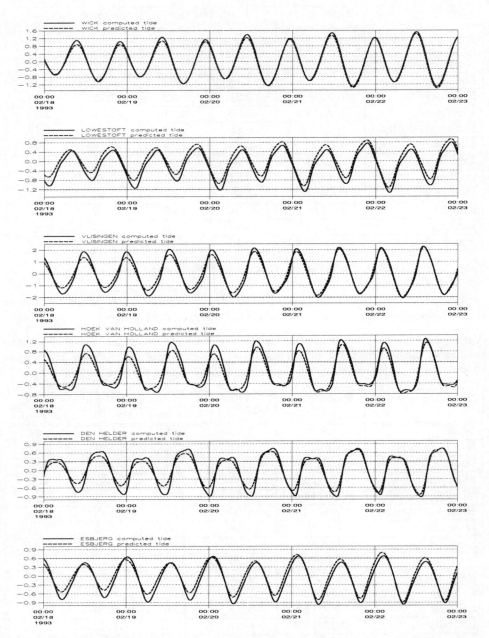

Figure 5.3: time series of predicted and computed tidal water levels at (from top to bottom): Wick, Lowestoft, Vlissingen, Hoek van Holland, Den Helder and Esbjerg

information about the storm entering the model through the Southern boundary condition. Moreover, significant errors are also present at the northern Dutch coast and specially along the western Danish coast. Within the last mentioned area, it is observed at Esbjerg that the main peak of the storm is calculated between two and three hours earlier and its magnitude is highly overestimated. Figure 5.4 shows the time series of the observed and modelled water levels at six different stations for the five-day period considered.

5.1.3   Filter set-up

For every test in this section, ten water level measurements have been assimilated from the 13 stations available. The other 3 have been kept as verification points in order to estimate if the filter is also improving the model results at different points than the assimilated measurements. The stations used for validation are: Hoek van Holland, Harlingen and Delfzijl. The standard deviation of the measurement noise has been considered equal to 0.05 m. at every station but Hanstholm, where the amplitude of the water level is smaller and therefore the standard deviation was set equal to 0.03 m.

It has to be observed that the filter performance is related to the definition of the open boundary conditions in the model set-up. Some of the model approaches for calculating the flux along the open boundary can generate instabilities in the filter. In this case, the boundaries have been defined only by water levels. In the case of inflow, the direction of the flux is unknown, therefore the flux is considered perpendicular to the boundary, with the component in the x-direction (along the boundary) equal to zero. In the opposite case (outflow), the component in the x-direction can be calculated as an extrapolation from the closest grid points, while the component in the y-direction is calculated from the boundary itself. In the transition between inflow and outflow the flux along the boundary can experience a sudden change in magnitude, from zero to a large value of viceversa. This discontinuity, generates instabilities in the filter solution that can propagate through the model. Due to the linearisations used in the RRSQRT filter and explained in chapter 3, the filter cannot deal with strong non-linearities as the one explained here. A twin test was carried out using the RRSQRT and the Ensemble Kalman filter defining the boundary condition as explained above. While the Ensemble Kalman filter provides a smooth solution in the areas close to the boundaries, the RRSQRT presents instabilities just after the discontinuity at the flux along the boundaries takes place. To avoid this problem, the flux along the boundary is set equal to zero for inflow as well as for outflow conditions for the RRSQRT filter, when the open boundary condition is defined only with water levels. Figure 5.5 shows the time series of the flux in the x-direction at the boundary and at the first internal point for two filter applications where the two possible boundary definitions explained above have been used. It is easy to see how the discontinuity grows in time.

The second step for the filter set-up has to do with the definition of the filter param-

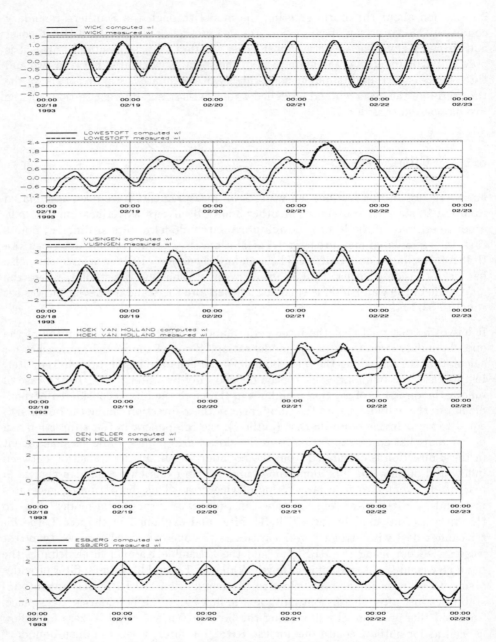

Figure 5.4: time series of observed and calculated (deterministic) water levels at (from top to bottom): Wick, Lowestoft, Vlissingen, Hoek van Holland, Den Helder and Esbjerg

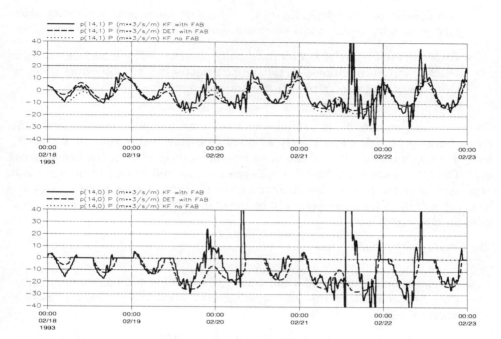

Figure 5.5: time series of the flux in the x-direction for a point at the southern boundary (14,0) and its northern neighbouring point (14,1), for the cases of a filter with flux along the boundary, a filter without flux along the boundary and a deterministic model with flux along the boundary.

eters. In the previous chapter we have studied the influence of certain parameters on the filter performance when incorrect error statistics are used. In this case we have knowledge of the error measured at the 13 observation points as the difference between the deterministic model and the measurements. The period of time as well as the number of stations in relation to the number of model grid points is not enough to estimate the real statistics of the errors in the system. At this point, it is necessary to make an assumption of the value of the filter parameters. Certain tuning or calibration is required, but the knowledge acquired from previous chapter will make it easier to fulfill this task. Moreover in this real application, the source of the errors is not completely known. Hence, it is necessary to assume the source of the errors. The possible sources are:

- *Errors in the momentum equation.* These errors are mainly associated with errors in the meteorological forcing terms, i.e. uncertainties in the wind velocity and atmospheric pressure.

- *Errors in the open boundary definition.* The model open boundaries are defined using tidal data and a correction term associated with changes in the atmospheric pressure (equation 5.1). Surges or water level variations which take place outside the model domain are not included in the boundary.

- *Errors in the surface elevation (continuity equation).* The model definition in terms of initial conditions, friction or bathymetry can contain errors which in this case can be associated with the surface elevation.

To obtain the best result as possible from the data assimilation application the errors have to be defined as closely as possible to the real errors. A sensitivity analysis using a combination of the three possible sources of error has been carried out. The filter parameters have been chosen from a combination of intuition and experience acquired from earlier chapters. Once the type of noise to be applied has been defined, the filter parameters will be tuned in order to obtain the best possible filter for the present case.

The filter's characteristics are:

- Interpolation between model and noise grid uses a local Kriging interpolation, the influence radius is the $\sqrt{2}\times$ (grid factor) and the correlation coefficient is equal to 0.98.

- the number of modes has been chosen equal to 100, to be sure that this parameter does not influence the result.

- A large grid factor equal to 10 has been chosen to avoid many heavy computations to define the noise in the momentum equation as well as the noise in the water levels.

- The temporal correlation has been defined using a first order autoregresive process as defined in section 3.1.2 with a lag-one autocorrelation coefficient equal to 0.9

- The noise in the momentum equation has been defined with a correlation coefficient of 0.8 and a standard deviation of 0.00025 m^2/s^2

- The boundary noise is specified independently for each boundary. The Northern boundary noise uses a correlation coefficient of 0.9, standard deviation of 0.05 m. and it is defined using a grid factor of 3, with respect to the model boundary. On the other hand, the southern boundary has the same parameters except the grid factor which is set to 1, i.e. it coincides with the model grid.

- The noise in the water levels is defined using a correlation coefficient of 0.8 and standard deviation of 0.02 m/s.

The first test to be carried out will define which noise type or combination of them, provides the best filter performance. The possible combinations of the three aforementioned error sources have been tested using the 100 leading eigenvalues ($M = 100$). Test results are presented in table 5.5 as well as the standard deviation of the error in the water level at all the stations and for the period of study considered. The name of the different tests indicates the kind of noise used: c and m represents noise in the boundaries or water levels and in the momentum equation respectively. The value associated with c is equal to 2 if the error in the water level field is considered, 1 if noise at the boundaries is used and 0 when none of the previous is applied. The value associated with m is equal to 1 if noise in the momentum equation is considered, and equal to zero if it is not.

Station	Determ. std (m.)	c2m1 std (m.)	c1m1 std (m.)	c0m1 std (m.)	c1m0 std (m.)	c2m0 std (m.)
Wick	0.28	0.07	0.07	0.12	0.08	0.08
North Shields	0.30	0.04	0.05	0.04	0.12	0.05
Lowestoft	0.28	0.13	0.15	0.12	0.21	0.18
Sheernes	0.63	0.17	0.25	0.33	0.32	0.23
Vlissingen	0.63	0.12	0.14	0.20	0.28	0.17
Hoek van Holland	0.36	0.25	0.15	0.16	0.16	0.28
IJmuiden	0.34	0.11	0.19	0.15	0.27	0.13
Den Helder	0.26	0.09	0.10	0.09	0.16	0.07
Harlingen	0.40	0.31	0.28	0.26	0.30	0.34
Huibertgat	0.37	0.08	0.16	0.12	0.31	0.10
Delfzijl	0.67	0.51	0.55	0.54	0.55	0.55
Esbjerg	0.39	0.06	0.13	0.08	0.27	0.07
Hantsholm	0.23	0.03	0.07	0.05	0.15	0.04

Table 5.5: Test of different error sources combination against the deterministic model

The first conclusion from the result of the test is that the best agreement between computed and observed water levels, is obtained when the noise is defined as a combination of noise in the momentum equation and noise either in the water levels or in boundary conditions. The best result at the observation points has been obtained from the test c2m1, while for the validation points it has been obtained from c1m1. This could indicate that the use of noise in the water levels generates a correlation between variables at different locations, which could be excessively strong.

Using the set-up c2m1, the influence of the number of modes has been tested. Results using a different number of modes are presented in table 5.6.

It can be observed from the results in table 5.6 that at measurement positions the filter results improve when increasing the number of leading eigenvalues. For $M = 25$ the filter produces very small corrections in the model. For values of M larger than 75, the improvement in the filter corrections slows down. Moreover, though the best result at the measurement positions is obtained using $M = 200$, the results at the validation positions slightly deteriorates. This effect is a consequence of the possible

Station	Determ. std (m.)	M=200 std (m.)	M=100 std (m.)	M=75 std (m.)	M=50 std (m.)	M=25 std (m.)
Wick	0.28	0.04	0.07	0.10	0.22	0.27
North Shields	0.30	0.02	0.04	0.05	0.12	0.25
Lowestoft	0.28	0.08	0.13	0.17	0.20	0.28
Sheernes	0.63	0.09	0.17	0.23	0.36	0.51
Vlissingen	0.63	0.07	0.12	0.13	0.29	0.49
Hoek van Holland	0.36	0.26	0.24	0.18	0.16	0.25
IJmuiden	0.34	0.08	0.11	0.12	0.15	0.26
Den Helder	0.26	0.05	0.09	0.12	0.15	0.24
Harlingen	0.40	0.32	0.31	0.29	0.27	0.38
Huibertgat	0.37	0.07	0.08	0.10	0.21	0.35
Delfzijl	0.67	0.65	0.51	0.49	0.50	0.60
Esbjerg	0.39	0.03	0.06	0.08	0.19	0.32
Hantsholm	0.23	0.04	0.03	0.04	0.09	0.17

Table 5.6: Test using c2m1 for different number of modes

increase on the correlation among different grid points due to the use of a larger M. With respect to computing time the time needed to compute the case with $M = 100$ is 1.5 times the time needed to run the case with $M = 75$. As it was introduced in the previous chapter, computing time grows very rapidly with the number of modes for the RRSQRT filter. Due to the important gained in computer time and because the result obtained with the filter for this value of M is very close to the one obtained using a larger M, a value of 75 has been selected. The filter with a rank approximation of 75, provides a solution close to convergence and simultaneously is not very demanding in computer resources.

With the previously selected filter (noise defined by c2m1 and $M = 75$) the sensitivity of the filter with respect to the other parameters have been tested. The filter parameters are: regression coefficient, correlation coefficient, standard deviation DMO, standard deviation of water levels and grid factor. The standard deviation of the error in water levels at the 13 stations obtained from five different tests, are shown in table 5.7.

The most sensitive parameter is the grid factor, i.e. the ratio between the noise grid and the model grid. Decreasing this parameter from 10 to 6 produce a small improvement mainly on the measurement positions. Simultaneously, the increase of the number of noise points makes the filter computationally more expensive because the eigenvalue decomposition of a larger matrix has to be calculated. In general, it can be concluded that the filter in this case is not very sensitive to small changes in the parameters. Small improvement can be achieved increasing the standard deviation of the noise at the momentum equation as well as the noise at the water levels. The test of all the possible combination of changes in all the parameters is an unthinkable task. Few combinations were tested and they did not present sig-

Station	Determ. std (m.)	$\alpha=95$ std (m.)	$\rho=0.9$ std (m.)	$\sigma_m=0.0004$ std (m.)	$\sigma_{wl}=0.04$ std (m.)	gfac=6 std (m.)
Wick	0.28	0.09	0.08	0.08	0.07	0.09
North Shields	0.30	0.05	0.05	0.05	0.05	0.05
Lowestoft	0.28	0.17	0.20	0.15	0.17	0.12
Sheernes	0.63	0.23	0.22	0.22	0.20	0.18
Vlissingen	0.63	0.13	0.13	0.12	0.14	0.12
Hoek van Holland	0.36	0.24	0.26	0.24	0.26	0.26
IJmuiden	0.34	0.11	0.13	0.12	0.12	0.12
Den Helder	0.26	0.10	0.10	0.10	0.11	0.08
Harlingen	0.40	0.31	0.29	0.29	0.33	0.32
Huibertgat	0.37	0.11	0.09	0.10	0.09	0.08
Delfzijl	0.67	0.49	0.55	0.51	0.48	0.51
Esbjerg	0.39	0.07	0.07	0.07	0.06	0.06
Hantsholm	0.23	0.04	0.04	0.04	0.03	0.03

Table 5.7: Sensitivity of the filter with noise type c2m1 and $M = 75$ to modifications in the filter parameters

nificant improvements in the filter results. Therefore, considering single parameter modification, the value of σ_m=0.0004 m^2/s^2 produces the best improvement (see table 5.7) and is going to be considered for the final filter set-up.

All the tests carried out during this section have shown that the most influential parameter is the rank of the square root approximation of number of leading eigenvalues M. The selected M will depend on computer requirement as well as the degree of accuracy required in the filter results. In a linear case, Verlaan [61] has shown than the larger the value of M, the more accurate the filter results.

The filter that has been chosen for the North Sea model test has the following features:

- $M = 75$

- Grid reduction factor $= 10$

- Lag-one autocorrelation coefficient 0.9

- The noise in the momentum equation has a correlation coefficient of 0.8 and a standard deviation of 0.00040 m^2/s^2

- The noise in the water levels has a correlation coefficient of 0.8 and standard deviation of 0.02 m/s.

5.1.4   Hindcast results

The hindcast results have been presented already because they correspond to the test '$\sigma = 0.0004$' included in table 5.7. Moreover it is important to have a visual representation of the filter corrections. Figures 5.6 and 5.7 show the comparison of the time series of water level obtained from the deterministic model, the observations and the filter application. The time series presented in figures 5.6 and 5.7 represent five measurement positions, which data have been assimilated into the model (Wick, Lowestoft, Vlissingen, Den Helder and Esbjerg) and one validation station (Hoek van Holland). It can be observed that the filter is providing good results not only at the measurement stations but also in the validation point. This conclusion could be easily reached from the values in table 5.7.

In order to observe the differences between the deterministic model and the filtered model results, the water level fields for both cases are shown in figure 5.8. This figure shows the output of the models at two different time steps (20/02/93 22:00 and 21/02/93 3:00). Important corrections can be observed in the model for example in the wave reaching the Danish coast which has been delayed and its amplitude

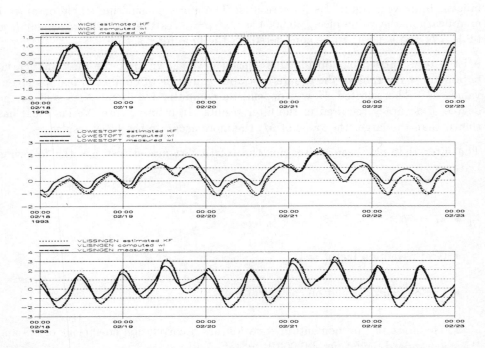

Figure 5.6: time series of water levels from the deterministic model, observations and estimated with the Kalman filter at (from top to bottom): Wick, Lowestoft, and Vlissingen

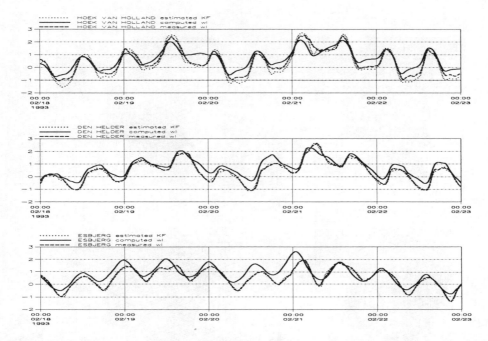

Figure 5.7: time series of water levels from the deterministic model, observations and estimated with the Kalman filter at (from top to bottom): Hoek van Holland, Den Helder and Esbjerg

decreased, providing a better agreement with the values measured in Esbjerg.

Moreover, the filter can provide some valuable information from different sources such as the Kalman gain, the error covariance matrix and the error variables contained in the augmented part of the state vector. These parameters provide information about the filter performance and behaviour. The Kalman gain for each measurement gives an idea of the influence of this measurement in other model areas. Figure 5.9 shows the Kalman gain for water levels at the 10 measurements assimilated at 00:00 on the 21/02/93.

It can be observed that the stations located at the southern part of the model have a very localised influence, while northern stations have a larger influence area. This feature is related with the main circulation that takes place in the North sea which consists on a Kelvin wave travelling anti-clockwise from the northern British coast. Moreover, the last measurement station, Hanstholm, influences a very large area. This is a consequence of using a smaller value of the measurements noise than for the other stations. It was previously mentioned that this station has a water level amplitude smaller than the other stations and therefore a smaller measurement noise was selected.

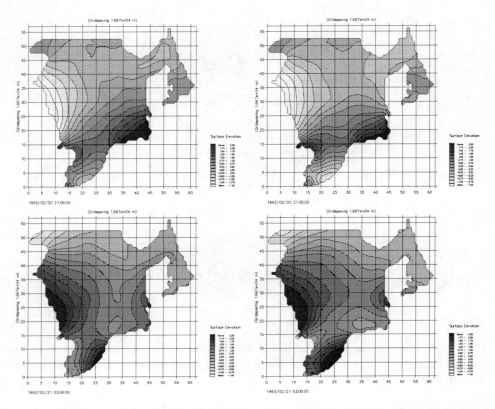

Figure 5.8: Water level fields obtained from the deterministic model (left) and the filtered model (right) at 20/02/93 21:00 (top) and 21/02/93 15:00 (bottom)

Another important source of information from the filter is the error covariance matrix. As introduced in chapter 4 it provides an estimation of the error in the system. The error covariance matrix is not explicitly calculated in the RRSQRT algorithm. An approximation of rank q of the square root of the error covariance matrix is calculated instead. In figure 5.10 the square root of the diagonal elements of the error covariance matrix, that represent the standard deviation of the water levels are shown for each grid point of the model at one specific time step.

The values of the standard deviation for each grid point are almost constant during the test. The filter estimates the largest error in areas where it is not correcting significantly, i.e. the central part of the model and the region between Norway, Sweden and Denmark, called the Skagerrak and Kattegat. The smallest values of the standard deviation are observed in areas close to the measurements and also in areas close to the open boundaries. The constant behaviour of the error estimation can be observed more clearly in figure 5.11, where time series of the standard deviation of the error in water levels at three locations are plotted.

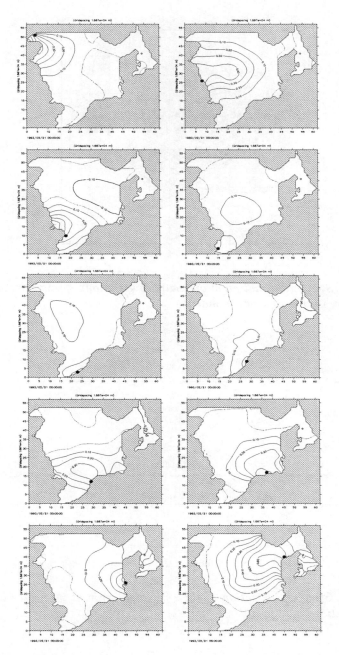

Figure 5.9: Kalman gain for the water elevations for the 10 measurement

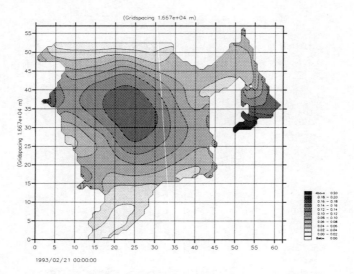

Figure 5.10: *Standard deviation of the water levels at one specific time step obtained from the filters error covariance matrix*

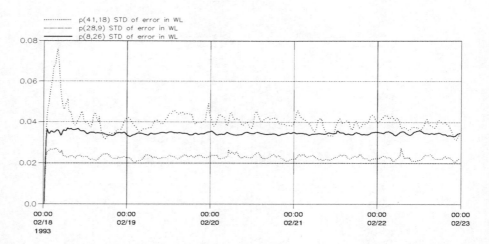

Figure 5.11: *Standard deviation of the estimated error at the water levels at three different locations*

Different behaviour is observed for the estimated values of the errors in the augmented part of the state vector. The use of coulored noise provide an estimation of the assumed error in the system. In this case errors were considered in the water levels and errors in the *DMO* terms. These errors are far from constant and in some of the stations it could be observed that the error is strongly related with the tide.

5.1.5 Forecast results

One important aplication of data assimilation is in the field of storm surge forecasting (see for example Heemink [31]). Typical applications assimilate data into the model up to the time of forecast. At this point, the best possible initial conditions for the deterministic forecast that are based on the data and the model have been generated. The error covariance matrix and the augmented part of the state vector (in the case of time-coloured noise) provide some information of the actual error in the system and how this error will propagate in the forecast period. When the estimated errors are systematic, i.e. they can be considered nearly constant in time, the model can be corrected with these values during the forecast period. However, the propagation of the system errors is usually defined using a simple model, for example a first order autoregresive process (see chapter 3). This process works as an exponential decay function of the latest estimated value of the error prior to the time of forecast. Therefore the forecast results will be strongly affected by how the simple propagation model can resemble the real error distribution in time. An initial test where the estimated model errors were propagated during the forecasting period showed that the forecast result not only did not improve but also for most cases it produced worse results than the deterministic model. This result as discussed it before is due to the use of simple models and correlation functions to define the error statistics. Due to the complexity of the model error forecast, pure model initialisation for forecasting has been used here. An example, focusing on the forecast of the first hours of the 21/02/93 is presented. Four different forecasts, initialising the model every three hours from 20/02/93 15:00 have been calculated. The forecast error is quantified by the RMSE between observed and forecasted surface elevations. Figure 5.12 shows the time series of water level for the different forecast initialisations at Esbjerg and Huibertgat compared to the deterministic model, the model using measurements during the five day period (hindcast) and the observations.

In order to quantify the improvement or deterioration in the forecast caused by the filter initialisation the following expression has been used

$$improvement(\%) = 100 \times \frac{RMSE(Deter) - RMSE(For)}{RMSE(Deter)} \qquad (5.2)$$

This parameter represents the improvement with respect to the error in the deterministic model. A negative improvement represents that the forecast with the deterministic model has smaller error than the filter initialised forecast. Table 5.8 shows the improvement in the forecast globally for all the stations and in the two stations presented in figure 5.12. The values of the root mean square error used in equation 5.2 have been averaged over 4 periods of 3 hours.

The performance of the filter forecast varies geographically in the model domain.

At the northern British coast the filter results deteriorate as compared to the deterministic model. The cause of this is that the filter had a very strong influence for correcting this area, and therefore the sudden change of conditions at the boundaries (from corrected to deterministic boundaries) can produce a discontinuity in

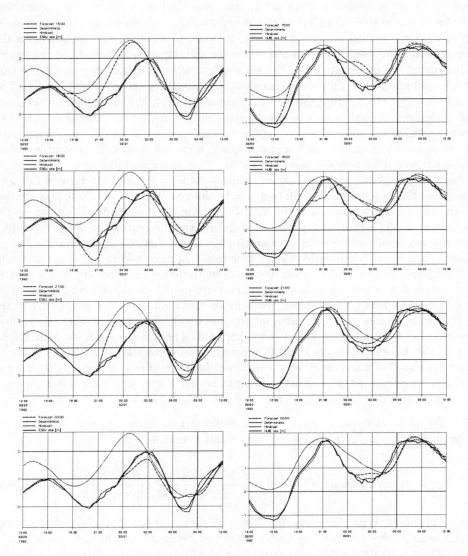

Figure 5.12: Time series of water levels: observations, hindcast, deterministic and forecasted. From top to bottom, 4 different initialisation times for the forecast from 15:00 20/02/93 to 00:00 21/02/93 every 3 hours. Results for two different stations: Esbjerg (left) and Huibertgat (right).

Lead time (h)	Average (%)	Esbjerg (%)	Huibertgat (%)
0-3	35	52	41
3-6	18	62	15
6-9	9	46	6
9-12	-12	10	5

Table 5.8: Improvements in the forecast with intial conditions obtained from the filter

the result. Close to the southern boundary, a considerable improvement is achieved during the first 6-9 hours. After this period of time the result deteriorates. At the northern Dutch coast the forecast produces significant improvements only during the first 6 hours, and mainly at the measurement stations. The best forecast for the longest lead-time is produced in Esbjerg, at the western Danish coast. A Kelvin wave that travels along the coast of the North sea, reaches this area after passing through the British and the Dutch coasts. Moreover, the northern boundary has a direct influence on this area. Both phenomena are corrected by the filter, therefore these corrections will reach the western Danish coast after some time. The last station, Hanstholm, presents a good forecast during the first 9 hours despite of the strong influence from the northern boundary.

5.2   Conclusions

In this chapter the implementation of the RRSQRT filter in a two-dimensional hydrodynamic model has been tested. The filter parameters have been calibrated in order to provide the best results. The most important parameter, as expected, is the rank of the square root approximation of the error covariance matrix (number of leading eigenvalues M). The computational time grows very rapidly with M. Therefore, this parameter has to be selected in order to provide good results for a reasonable associated computer cost.

Another important fact to be considered is the definition of the system noise. Better results have been obtained when a mixed noise (boundary and momentum, or water levels and momentum) has been defined. The system error can be better described and consequently the model can be corrected faster and more accurately. For hindcasting the filter can force the model towards the measurements and therefore the global model result is considerably improved.

In the previous chapter the noise at the boundary and at the meteorological forcing were quite accurately estimated for the twin tests. In that case, the noise was represented by a simple AR(1) model and this information was given to the filter. In a real case the error structure and error statistics could be represented by very complex functions. Also in the North Sea model, the RRSQRT filter used the simple AR(1) model to propagate the error. Due to this simplistic approach the model errors cannot be estimated correctly and the error covariance matrix becomes nearly constant. A constant error covariance matrix leads to a constant Kalman gain matrix. However, very similar filter results could be obtained using this nearly constant Kalman gain for future state updates, as in a steady state filter application.

For forecasting, because the estimated errors are not correctly estimated, they cannot be propagated forward and used to correct the model forecast. The RRSQRT filter in this case can be used for initialisation of the model prior to forecasting, i.e. the filter is applied while measurements are available and then a deterministic forecast is issued using the best possible initial conditions. This solution provides good results, up to 9-12 hours in some areas of the model.

In order to improve the filter performance for hindcasting as well as for forecasting, a better definition of the error structure is necessary. Based on all the measurement available, the temporal correlation of the error has to be defined with more complicated expressions. Moreover, the error covariance matrix of the system noise, which here is considered constant in space, has to represent more accurately the differences that are presented in nature among the different model areas. The application of very sophisticated data assimilation techniques with a simple definition of the error structure and error statistics has an important limitation in model corrections as well as forecasting capabilities.

Finally, it is important to notice that when a non-linear definition of the open boundary has been used, the filter has exhibited instabilities. This result was expected since the extended Kalman filter and therefore the RRSQRT filter are very well suited for linear and weakly non-linear model, but they could fail for strong non-linearities (see Evensen [21]).

Chapter 6

Application on a 2D Nested hydrodynamic model. Test case

In storm surge models as well as in operational warning systems, time is a very important factor. Complicated bathymetries need to be described using small grids, with the associated increase in computational time. Nested models allow one to have a better resolution in desired areas while maintaining a coarser grid in the rest of the model. Nested areas are solved simultaneously with the rest of the model. An application of a nested model can be found in Vested *et al.* [65].

Data assimilation algorithms need to run the hydrodynamic model several times, and store a large amount of variables. The computational burden is proportional to the number of model variables. Large models can become prohibitive. Application of data assimilation in large models which need a good resolution in specific areas can be feasible with a nested model.

6.1   Special features of the Nested model

The nested hydrodynamical version of the MIKE 21 model (see chapter 2), solves the hydrodynamic equations simultaneously in a number of dynamically nested grids. An important difference between nesting and boundary transfer from a coarse model to a finer one is that in nesting the information between grids travel in two directions, i.e. from the coarse to the fine grid and viceversa. On the other hand, in a boundary transfer case information travel only from the coarse to the fine grid. Both techniques are sometimes called two ways and one way nesting simultaneously.

In order to ensure model stability and smooth transition between areas, some limitations have to be considered. The most important are:

- Open boundaries can only been defined in the coarsest grid

- The spatial resolution from one level to another is reduced by a fixed factor, which is equal to 3.

- The water depths in common grid points along the internal boundaries should be equal in both the coarse and the fine grid. Between the common points along the internal boundary, the water depths in the fine grid are linearly interpolated using the values at the common points.

- The water depth in the coarse grid has to be equal in three points orthogonal to the internal boundary (at the border and one point at each side). Therefore the first four points orthogonal to the internal boundary in the fine grid have the same water depth. The intention of these corrections is to avoid instabilities in the internal boundaries.

More details about the model are included in DHI [19].

6.2   Implementation of the RRSQRT filter

The state of the system in this case is represented by the water levels and flux densities at every grid point as in the standard two-dimensional case. It is necessary to consider that now the grid is not regular along the model domain, but there exist areas with finer grid definition. The state vector that contains all the variables represented in a domain with fine and coarse grid definition is defined by x^{tot}. The state vector that represents the same domain as the former, but in this case all the variables are represented in a coarse grid resolution is defined by x^{coar}. These two vectors are related by the expression:

$$x_k^{tot} = \bar{\Lambda} x_k^{coar} \qquad (6.1)$$

where $\bar{\Lambda}$ contains the interpolation weights between the variables in x^{tot} and x^{coar}.

The main sources of errors that are normally considered in a hydrodynamical model are the open boundary conditions and the wind field. The open boundary conditions in the nested model are only defined in the main area (coarsest area). Therefore, the same implementation used for the standard two-dimensional model can be applied here. The meteorological forcing is normally obtained from models that are defined in coarser grids than the model grid itself. The definition of the errors in the meteorological forcing in a coarse grid as it was applied in the standard two-dimensional model, can also be considered in the nested implementation.

The number of water points and consequently the number of variables, increases very rapidly in a nested application. Usually, the fine grid areas (detail areas) contain more points than the coarser grid. In order to illustrate this a model with a grid of 30 × 30 points is considered, containing a finer grid with the same number of grid points which contains a third grid inside with the same number of grid points.

The three models have 800, 800, and 900 water points respectively. Considering the variables water level, momentum flux in x-direction at time t and momentum fluxes in y-direction at times $t + \frac{1}{2}$ and $t - \frac{1}{2}$, the state vector for this model will contain approximately 10,000 variables, while the state vector corresponding to the main area will contain 3,600 variables.

This implementation of data assimilation into a nested model will have the following features

- The model is propagated using the complete state vector x^{tot}

$$x_{k+1}^{tot,f} = f(x_k^{tot,a}, u_k) \qquad (6.2)$$

- The error covariance matrix is represented in the coarsest grid. Therefore the number of variables in the state vector is the same as if only the coarsest grid is considered.

- The error covariance matrix is propagated using the complete nested model, i.e. the error is interpolated from the main grid to the internal grids.

$$S_{k+1}^{*f,coar} = \left[\bar{\Lambda}^{-1} \left(F_k \bar{\Lambda} S_k^{a,coar} \right) \quad \Big| \quad \Lambda_k [Q^{coar}]_k^{1/2} \right] \qquad (6.3)$$

- If the measurement is located on a grid point of an internal area, in order to calculate the innovation, the value is extrapolated to the main area. In the case of a model with only one area the vector H consisted of zeros everywhere and 1 at the measurement position. In the case of more than one area this vector will have non-zero values at the positions that correspond to the four points of the coarse grid that surround the measurement point. Figure 6.2 shows the measurement p located on the finer grid, and the surrounding points of the coarser grid, $P1$, $P2$, $P3$ and $P4$.

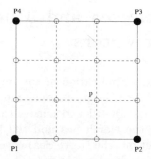

Figure 6.1: Points used for grid interpolation from the fine to the coarse grid

The error covariance matrix contains only information about the coarsest grid, therefore the measurement p must be related to the available information. Vector H will represent the relation between the position p and the positions $P1$, $P2$, $P3$ and $P4$ in the coarsest grid. It will have zeros everywhere but at the correspondent positions of $P1$, $P2$, $P3$ and $P4$.

$$H_i = \begin{bmatrix} 0 & \dots & 0 & \alpha_1 & 0 & \dots & 0 & \alpha_2 & 0 & \dots & 0 & \alpha_3 & 0 & \dots & 0 & \alpha_4 & 0 & \dots & 0 \end{bmatrix} \tag{6.4}$$

The relation can also be expressed as:

$$p = \alpha_1 P_1 + \alpha_2 P_2 + \alpha_3 P_3 + \alpha_4 P_4 \ \ \text{with} \ \sum_{i=1}^{4} \alpha_i = 1 \tag{6.5}$$

- The values of the gain correspond to variables in the main area and therefore they have to be interpolated to the secondary areas. Equation 6.6 is the new updating equation for this case

$$\left(x_{k+1}^{tot}\right)^a = \left(x_{k+1}^{tot}\right)^f - \begin{pmatrix} \bar{\Lambda} & K_k^i \end{pmatrix} \ (Error)^i \tag{6.6}$$

where the Kalman gain, that is defined in the coarse state vector has to be converted to the complete x^{tot} using the interpolation matrix $\bar{\Lambda}$.

Under these assumptions, the associated cost of the data assimilation in the nested model is comparable to the standard two-dimensional model in matrix algebra, storage, number of modes and eigenvalue calculation. The main difference is that the time associated with a run of the nested model is larger than the one in the standard model.

6.3   A test case: a regional model of the North sea and Baltic sea and a nested area in the Danish Inner waters

The aim of this test case is the application of the implemented data assimilation method, the RRSQRT filter, to a large regional model with important physical differences among different areas. In order to achieve this, the filter has been applied to a regional model covering the North Sea and the Baltic Sea. The area of the model used in the previous section (the North Sea) covers approximately half of this regional model. The model's bathymetry is defined with a grid size of 9 nautical miles (16670 m.) in both directions. At the two open boundaries located at the North Sea, water levels are specified using the same procedure explained for the

North Sea model. For the simulation period, wind velocity fields and pressure fields are available every three hours and they are linearly interpolated at every model time step (set equal to 10 min.). The flow resistance for this case is defined with a constant Manning number equal to 32 $m^{1/3}/s$. The model is initialised the 01/10/97 at 00:00 with water level and velocity fields obtained from a spin-up simulation of 48 hours.

The use of regional models with finer bathymetries in specific areas of interest is a common practice. In this particular test case, a nested area of the Danish inner waters has been defined in order to obtained a better detail of the water level and current fields in this specific zone. This local model is defined in a grid with origin in (48,18) of the coarse grid, and a grid size of 3 nautical miles (5667 m.) i.e. a third of the grid size of the regional model.

The performance of the implementation of the RRSQRT filter in both, the standard hydrodynamic model (HD) and the nested hydrodynamic model (NHD), is tested in the regional model. During the assimilation procedure water level data from 14 stations spatially distributed along the model are assimilated and the results area validated using another 7 available water level stations. The result could provide a good estimate of the global performance of the filter, i.e. the filter not only improves the model results at the measurement position but instead it improves the global model result. Table 6.1 presents the position of the measurement and validation points in the regional and local grids.

Figure 6.2 shows the model bathymetry used in the hydrodynamic model including the position of the measurement and validation stations in the model domain. Moreover, the second set-up when a local bathymetry is used in the nested model is also presented, including the measurement and validation stations.

A short period of data was available for this study. The period covers from 01/10/97 00:00 to 03/10/97 00:00, when continuos measurement of water levels were available at the 21 stations aforementioned. The filter parameters used for this model vary slightly from the parameters applied in the North Sea model presented in the previous chapter. From a rapid sensitivity test they proven to provide good results. The parameters are:

- $M = 100$

- Grid reduction factor $= 8$

- Lag-one autocorrelation coefficient 0.97

- The noise in the momentum equation has been defined with a correlation coefficient of 0.9 and a standard deviation variable in space. The magnitude of the standard deviation varies from 0.0005 m^2/s^2 in the North sea to 0.0001 m^2/s^2 in the Baltic sea.

- The Northern boundary noise uses a correlation coefficient of 0.95, standard deviation of 0.1 m. and it is defined using a grid factor of 3, with respect to

Station	Regional x co-ordinate	Regional y co-ordinate	Local x co-ordinate	Local y co-ordinate
Aberdeen	7	43	-	-
Lowestoft	18	10	-	-
Esbjerg	44	28	-	-
Hanstholm	45	40	-	-
Frederikshavn	53	42	-	-
Goteborg	56	44	-	-
Hornbaek	59	33	34	47
Fredercia	50	30	6	36
Korsor	54	28	19	31
Gedser	58	23	30	15
Klagshamn	61	29	40	34
Tejn	68	28	-	-
Kungsholmfort	71	34	-	-
Marviken	75	51	-	-
Hvide Sande	43	32	-	-
Torsminde	43	35	-	-
Thyboron	43	37	-	-
Hirsthals	50	43	-	-
AArhus	52	33	12	45
Kobenhavn	61	30	37	39
Rodby	56	23	24	15

Table 6.1: Grid co-ordinates of the measurement and validation positions in the regional and local models.

the model boundary. On the other hand, the southern boundary has the same parameters except the grid factor which is set to 1, i.e. it coincides with the model grid.

- Measurements where available every 20 minutes, therefore the updating step of the RRSQRT filter takes place only every second time step.

In order to evaluated the performance of the filter the root mean square error (RMSE) between the observed and updated water levels are calculated and compared with the RMSE between the observed and the deterministic model simulation for both model, the hydrodynamic and the nested hydrodynamic. The RMSE has been calculated using the last 36 hours of simulation in order to do not take into account the filter initialisation. Table 6.2 presents the RMSE for the measurements and the validation stations obtained from the deterministic and the updated hydrodynamic and nested hydrodynamic models.

Both deterministic models yield, basically the same results. The use of the detailed

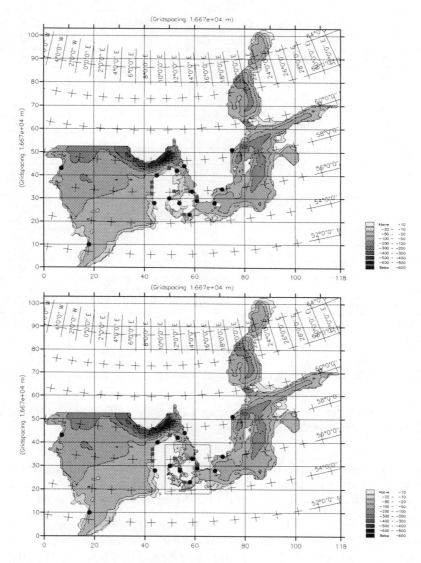

Figure 6.2: Top: North Sea and Baltic Sea model bathymetry (Depth in meters) with only one area. Bottom: North Sea and Baltic Sea model bathymetry (Depth in meters) with internal nested area. In both figures water level stations are represented with circles (measurements) and squares(validation).

area improves the result only in some of the stations while providing worse results in the others. This is a direct consequence of a poor schematisation of the area between Denmark and Sweden in the finer grid, resulting in worse results at the

Station	Deterministic RMSE (m.) HD	Kalman filter RMSE (m.) HD	Deterministic RMSE (m.) NHD	Kalman filter RMSE (m.) NHD
Aberdeen	0.43	0.14	0.43	0.15
Lowestoft	0.25	0.13	0.25	0.13
Esbjerg	0.43	0.11	0.43	0.12
Hanstholm	0.22	0.09	0.22	0.08
Frederikshavn	0.23	0.07	0.23	0.06
Goteborg	0.23	0.07	0.23	0.06
Hornbaek	0.28	0.06	0.26	0.05
Fredercia	0.21	0.05	0.19	0.06
Korsor	0.23	0.05	0.19	0.06
Gedser	0.17	0.07	0.25	0.08
Klagshamn	0.10	0.05	0.10	0.09
Tejn	0.17	0.07	0.17	0.08
Kungsholmfort	0.05	0.03	0.05	0.04
Marviken	0.03	0.02	0.03	0.02
Hvide Sande	0.23	0.13	0.23	0.12
Torsminde	0.25	0.10	0.25	0.11
Thyberon	0.28	0.12	0.28	0.12
Hirsthals	0.22	0.12	0.22	0.13
AArhus	0.22	0.14	0.22	0.17
Kobenhavn	0.27	0.12	0.27	0.16
Rodby	0.22	0.14	0.22	0.08

Table 6.2: RMSE for the deterministic and the updated hydrodynamic (HD) and nested hydrodynamic (NHD) models

stations located at the southern part of the fine grid (Gedser and Klagshamn). The results at the other stations located in the detail area are improved using the finer resolution. No further efforts for calibrating the models have been done.

The Kalman filter, for the HD as well as for the NHD case, corrects sufficiently the water levels in all parts of the regional model. At measurement points a global value of the RMSE equal to $0.072\ m.$ for the HD and $0.077\ m.$ for the NHD have been obtained which is only slightly higher than the assumed standard deviation of the measurement noise ($0.05\ m.$). At validation points the corrections are not that significant but still present a marked improvement (about 50% reduction of the RMSE) in all regions when compared with the deterministic model. Table 6.3 shows the global values of the RMSE errors for the deterministic and the updated HD and NHD models.

In general the results obtained in this case, in terms of reducing errors in water levels, are not improved by using a finer grid. Because the state vector is defined in the coarse grid, corrections in the fine area are obtained interpolating the error estimated

Station	Deterministic RMSE (m.) HD	Kalman filter RMSE (m.) HD	Deterministic RMSE (m.) NHD	Kalman filter RMSE (m.) NHD
Measurements	0.218	0.072	0.216	0.077
Validation	0.240	0.126	0.240	0.127

Table 6.3: Global RMSE values for the Deterministic and the updated hydrodynamic and nested hydrodynamic models

by the filter in the coarse area. The introduction of this new approximation causes a slightly worse performance of the filter in the finer grid. Simultaneously, the use of finer resolution provides a more detailed corrected velocity field as it can be easily observed when figures 6.3 and 6.4 are compared.

In order to show the differences between deterministic and updated results the water level and velocity fields at 19:00 on the 10/02/1997 obtained from the HD model are presented in figure 6.3. Important differences in the velocity field can be observed in this figure. The flux entering the Danish waters from the North Sea is not well represented in the deterministic solution. The updated model presents higher water levels entering from the Baltic sea than in the deterministic result. In general, the updated model is able to reproduce the high water level elevation at the Danish waters and the northern coasts of Germany and Poland, while this effect is smaller in the deterministic solution. Larger currents are obtained in the updated model, specially in areas with strong water level gradients, when using the fine resolution area, because in this case the velocity field is better represented. This result is presented in figure 6.4, where water level and velocity fields in the detailed area at 19:00 the 10/02/1997 are shown. This type of global improvement presented here is mantained through the whole simulation.

6.4   Conclusions

The RRSQRT filter has been implemented in a model that computes simultaneously the shallow water equations in different areas with different resolutions. The use of a finer resolution grid in some areas of the model yields better deterministic model results in these areas, when the model has been correctly calibrated.

The use of a finer resolution increases rapidly the number of computational points of the model. The state vector has been defined on the main (coarsest) grid with the aim of reducing the computational time and the storage requirements. This approximation, though providing a cheaper solution for the data assimilation application, did not improve the corrections of water levels in the detailed area, when compared with the ones obtained using only one area. More accurate results are expected if all the variables are defined in the state vector, because the interpolation of the Kalman gain from the coarse to the fine grid is not necessary in this case.

The approximation used in this implementation will probably fail if a third level of resolution area (a finer grid inside the actual fine grid) is used. In this case only global corrections could be achieved, because the noise is defined in the coarsest area. Detailed corrections could not be expected.

In cases where the main interest is focused in the water levels, as in the case of storm surge modelling, the implementation using only one area of the regional model can provide good enough estimation of the water levels.

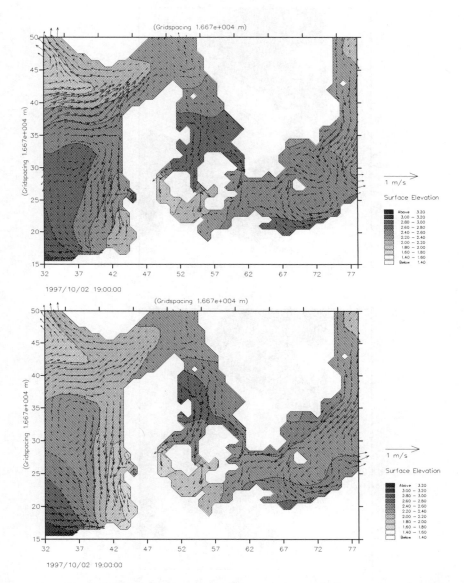

Figure 6.3: Water level and velocity field for the inner Danish waters at 19:00 the 02/10/97 calculated from the deterministic model (bottom) and the Kalman filter (top) using the hydrodynamic model (HD).

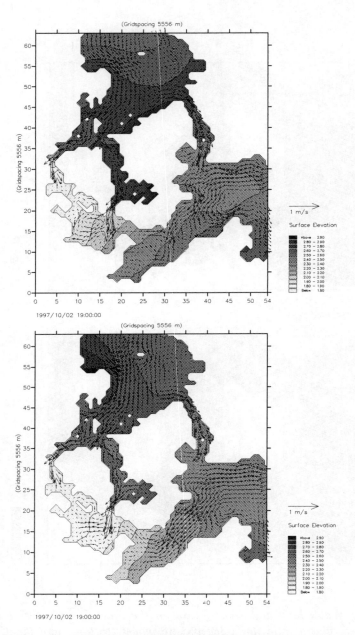

Figure 6.4: Water level and velocity field for the detailed area of the inner Danish waters at 19:00 the 02/10/97 calculated from the deterministic model (bottom) and the Kalman filter (top) using the nested hydrodynamic model (NHD).

Chapter 7

Initial tests on a 3D hydrostatic model

The number of applications of three-dimensional models in the field of regional coastal models is becoming more important nowadays. Hence, it is a natural step forward to apply data assimilation techniques on these type of models. For this purpose, a recently developed 3D hydrostatic model (Pietrazk *et al.* [51]), which uses the same solution algorithm as MIKE 21 to solve the continuity and momentum equations for the depth integrated variables, has been selected.

This chapter studies the application of the RRSQRT filter on MIKE 3 HS and outlines the similitude and differences between this application and the one carried out in the two-dimensional model. Moreover, results from a twin-test similar to the one presented in chapter 4 are also presented in order to discuss the filter performance. In this case, systematic errors at the open boundaries as well as in the meteorological forcings are considered.

7.1 Implementation in MIKE 3 HDHS

MIKE 3 HS is a three-dimensional hydrostatic model that solves the governing equation under the assumption of hydrostatic pressure. This implies that the vertical velocity can be obtained from the layer integrated continuity equation and therefore the vertical momentum equation is not needed. Another specific characteristic of the model is its vertical grid. The vertical grid is flow adaptive and follows the bottom contours. It is very well suited for weakly varying bathymetries. The model equations together with the numerical solution were introduced in chapter 2.

Based on the previous implementations in the depth integrated model MIKE 21, a similar procedure can be applied to the MIKE 3 HS model. The fact that MIKE 3 HS uses the same solution algorithm as MIKE 21 for the depth integrated variables, provides the opportunity of using a similar data assimilation technique to the one applied in MIKE 21. The error covariance matrix is only defined for the depth integrated variables ζ, p and q as in the two-dimensional case, i.e. it contains information about the error on these variables. In order to propagate the error covariance matrix, a first order finite difference approximation similar to the one presented in equation 3.24 is used. In this case the state vector contains all the model variables, and the error covariance matrix is only associated with the depth integrated variables. The expression in this case reads

$$\frac{\partial f}{\partial x} s_{j,k}^{DI,a} \approx \frac{f(x_k^{3D,a} + \epsilon \Lambda^{3D} s_{j,k}^{DI,a}, u_k) - f(x_k^{3D}, u_k)}{\epsilon} \tag{7.1}$$

where the x^{3D} is the state vector containing all the model variables, s_j^{DI} is the $j^t h$ column of the square root of the error covariance matrix representing the depth integrated variables, and the operator Λ^{3D} represents the distribution of the depth integrated values of s_j^{DI} to the layer integrated variables and the vertical velocities.

Moreover the Kalman gain provides the corrections to be made on the depth integrated variables. The updating equation can be written as

$$\left(x_{k+1}^{3D}\right)^a = \left(x_{k+1}^{3D}\right)^f - \left(\Lambda^{3D} \quad K_k^{DI,i}\right) \quad (Error)^i \tag{7.2}$$

Based on the corrections made on the depth integrated variables, the layer integrated variables are also corrected. The adopted method considers that the distribution of the corrections through the layers is a fuction of the ratio, for each layer, between the layer width and the total depth. Once the horizontal layer integrated fluxes have been obtained the layer integrated continuity equation is solved and the vertical velocities calculated. These steps in order to correct the complete model are represented in equation 7.2 by the matrix Λ^{3D}.

7.2   Twin test

In order to evaluate the performance of the data assimilation scheme the same synthetic test (twin test) used in chapter 4 for the application in the two-dimensional model, has been carried out. A hypothetical bay with an open northern boundary, a grid of 21 x 20 points, and a grid size of 10 x 10 km^2 has been used while the vertical dimension is discretised with 5 layers. The model bathymetry is shown in figure 4.1. The flow is forced by a sinusoidal variation of the surface elevation at the open northern boundary with a period of 12 hours and an amplitude range of 2 m. Meteorological forcing is included using wind and pressure fields from an artificially generated moving cyclone that moves in a west-east direction with a speed of 8.33 km/h (see figure 4.1). Physically, the main flow describes a Kelvin wave moving anti-clockwise in the bay region. From the result of this model, which is going to be addressed as the true model, time series of surface elevations have been extracted at three different positions. These time series are going to be used as the measurements to be assimilated in the wrong model. The measurement positions have been defined in horizontal co-ordinates at (1,16), (8,1) and (20,12).

7.2.1   Test 1: Systematic error at the open boundary

The three water level measurements are assimilated in a twin model with the difference that the open boundary condition has a phase error of one hour with respect to the original. The aim of the test is to recover the true model from the three measurements when the wrong open boundary is applied.

The filter parameters are the same used in chapter 4 for the case of error at the open boundary. The noise introduced in the boundary is temporarily correlated using a first order autorregresive process with a lag-one autocorrelation coefficient of 0.9. The residual errors are generated from a Gaussian distribution with zero mean and with a covariance matrix calculated from an exponential distance correlation model with a correlation coefficient of 0.9 and considering a standard deviation of 0.1 m. The rank reduction used in the RRSQRT filter equal to 20 ($M = 20$, as defined in previous chapters). The results from the three different models are presented and also compared in this section. These models are the true model, which has been already defined, the wrong model i.e. a model using the same specifications as the true model except that it uses the wrong boundary condition, and the model corrected by the filter. As explained before surface elevations obtained from the true model have been assimilated at (1,16), (8,1) and (20,12). These measurements have not been corrupted by noise, though measurement noise with standard deviation of 0.05 m. has been assumed. The surface elevation obtained with the filter at the three measurement positions are in good agreement the assimilated data as observed in figure 7.1 .

This figure presents the time series of surface elevations at the three measurement positions, obtained from the three aforementioned models (the true, the wrong and the corrected). Moreover, it is even more important to learn about the filter corrections in other points as well as for other kinds of different variables to the assimilated. Because we are dealing with a twin test, we have knowledge of the complete true model (it never happens in a real application). In order to verify the filter performance for surface elevation at other locations, three validation points have been selected: (5,15), (10,10) and (15,10). Figure 7.2 shows the time series of surface elevations at these locations. From these results it is easy to conclude that the filter is not only correcting the error at the observation points but in the overall model. Moreover, as in the case presented in chapter 4 for the 2D model, the systematic error at the boundary is quite accurately estimated and consequently improving the overall model result.

The correlations among errors for different variables are defined in the error covariance matrix. These correlations are hydrodynamically consistent because they have been obtained by propagating the errors using the hydrodynamic model itself. In this case the currents are generated by barotropic forces therefore the correction in the surface elevation and the boundary produces a significant improvement of the horizontal velocity field, as well as in the vertical.

Figure 7.3 presents the velocity profiles for all the points at $x = 4$ obtained from the three model at 01/01/90 15:00. It can be observed that the filter is able to modify the profiles providing an almost identical result to the true model. Small differences are still presented in areas close to the open boundary.

Finally, the profiles of the vertical velocity and the y-component of the velocity at

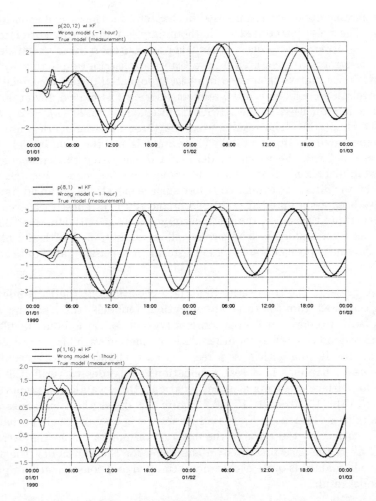

Figure 7.1: Time series of water levels from the true, the wrong and the updated model at the three measurement locations for Test 1.

one validation point, in this case (4,6) are presented in figure 7.4. The filter results almost match the true model in this case. It has been observed that the vertical velocity is corrected gradually during the simulation, i.e. the error after two days of simulation is very small, while after one day of simulation the error was still significant (on the order of half of the total error).

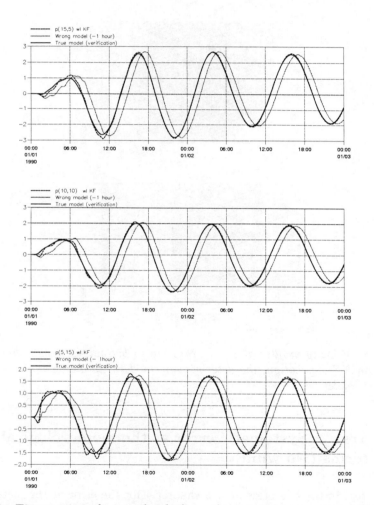

Figure 7.2: Time series of water levels from the true, the wrong and the updated model at the three validation locations for Test 1.

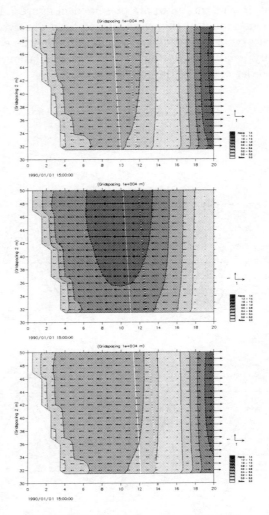

Figure 7.3: Velocity profiles at j=4. From top to bottom the results have been obtained from KF application, the deterministic model with the delayed boundary and the correct model.

7.2.2   Test 2: Systematic error at the meteorological forcing terms (No wind)

A similar test to the one presented in chapter 4 for the error at the meteorological forcing terms in the 2D model is repeated here for the MIKE 3 HS case. The aim is to test the filter under the existence of systematic errors in the meteorological forcing terms. The test consists of neglecting the meteorological forcing terms, i.e.

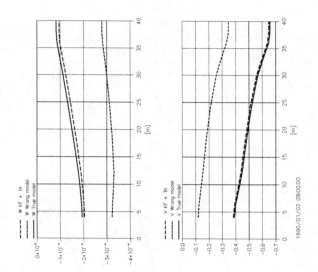

Figure 7.4: Velocity profiles at (4,6) for the vertical velocity (left) and the y-component of the horizontal velocity (right). The profiles have been obtained from the true, the wrong and the corrected model. The vertical scale for the vertical velocity is presented every 10^{-1} m/s and for the horizontal velocity every 0.1 m/s.

no meteorological input is given to the model. The same measurements obtained from the true model and used in the previous section have been used here. The filter parameters are very similar to those used in chapter 4. The noise introduced in the depth averaged *DMO* terms is temporarily correlated using a first order autorregresive process with a lag-one autocorrelation coefficient of 0.97. The residual errors are generated from a Gaussian distribution with zero mean and with a covariance matrix calculated from an exponential distance correlation model with a correlation coefficient of 0.98 and standard deviation of 0.1. The rank reduction used in the RRSQRT filter is equal to 50. The noise has been introduced on a coarser grid with a reduction factor of 4 with respect to the model grid. In order to study the filter performance, similar plots as in the boundary case are going to be presented. With regards to the surface elevations, the filter provides good corrections in the measurements as well as in the verification points. It can be observed from figure 7.5 that the error in water levels, which is generated by neglecting the meteorological forcing, is not very important when compared to the tidal amplitude. Figure 7.5 shows the water level obtained from the true, the wrong and the corrected model at two measurement positions (1,16) and (8,1) and at one validation point (15,5).

The system is mainly forced by the open boundary. Despite this, the differences between the models with and without the meteorological forces are significant. These differences are very important in other variables such as velocities. The filter is able to make corrections in the three components of the velocity. As expected the results

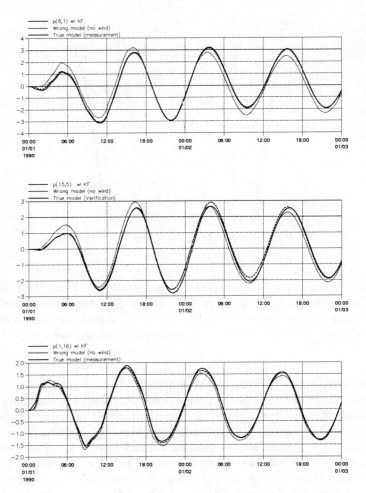

Figure 7.5: Time series of water levels at two measurement positions (top and bottom) and a validation point (centre), obtained from the true, the wrong and the corrected model.

are not as good as in the previous test, though the error in the velocity profile is reduced for all the components. For consistency, the velocity profiles at the same location (4,6) than for the previous test are presented in figure 7.6, and only at one specific time step. In general we can say that the filter produces better corrections in the vertical velocity (W) and in the x-component of the horizontal velocity (U) (the two smallest in magnitude), while the y-component (V) presents significant corrections only at the end of the two-day simulation. Increasing the standard deviation of the noise in the DMO terms, the corrections in water levels and in V are improved while in U and W worse results are obtained (too strong corrections).

Despite the improvement in the velocity profiles obtained with the Kalman filter, the shape of the profile is not significantly modified probably due to the simple approach used for corrections in the different layers. Because the correction in the depth averaged fluxes is just averaged through all the layers, the shape of the velocity profile is almost conserved. It has to be considered that this test is a very special and strong case, because the vertical distribution generated from the wind is not presented in the model when no meteorological forcing is used.

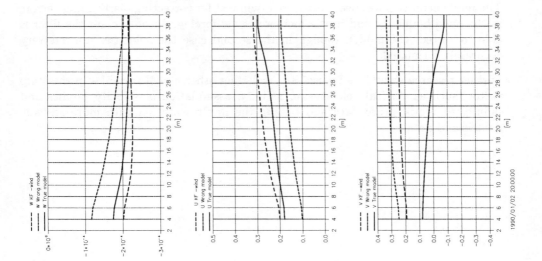

Figure 7.6: Velocity profiles at (4,6) for the vertical velocity (left) the x-component of the horizontal velocity (centre) and the y-component of the horizontal velocity (right). The profiles have been obtained from the true, the wrong and the corrected model. The vertical scale for the vertical velocity is presented every 10^{-1} m/s and for the two components of the horizontal velocity every 0.1 m/s.

7.3 Conclusions

The RRSQRT filter has been implemented into a three-dimensional model, which solves the governing equations under the assumption of hydrostatic pressure. The state vector in this initial implementation only contains the water levels and the two components of the depth averaged velocity, though the error covariance matrix is propagated using the full model equations.

The filter provides corrections for the water levels and the depth averaged velocities. These corrections are also used for correcting the velocity profile. All the modifications in the vertical velocity, as well as layer integrated velocities are proportional to the corrections in the surface elevation and the depth averaged velocities. In order to obtain correlations among all the model variables, they have to be included in the state vector.

Under the conditions of this implementation, baroclinic effects cannot be corrected. Moreover, if measurements of velocity profiles are available, they cannot be assimilated directly by the filter because only the depth averaged velocities are included in the state vector.

The implementation has been tested in a twin test for correcting biased model errors in the open boundary and in the meteorological forcing. In both cases the filter is able to correct the model, though in the second case the velocities are not very accurately estimated.

Further research should be focused to: including other model variables in the state vector, such as vertical velocities or salinity, assimilation of velocity profiles and salinity or temperature data and testing the implementation with more realistic models and data.

Chapter 8

Summary and conclusions

The aim of the research presented in this thesis was to integrate and further develop existing computationally efficient data assimilation techniques for large scale regional coastal models. The use of different numerical models during the coarse of this work made it necessary to develop specific data assimilation methods for each of the models.

The first part of this work deals with the description of Kalman filter techniques. In order to define the Kalman filter equations, the linear Kalman filter as well as the full extended Kalman filter are introduced. In addition, the definition of the Kalman filter using an augmented state vector formulation has been included. This formulation makes the filter more robust, specially in the case of biased errors.

Two suboptimal schemes have been described. The first scheme is a reduced rank square root (RRSQRT) approximation of the extended Kalman filter, which approximates the error covariance matrix by one of a lower rank. The second is the ensemble Kalman filter (EnKF), which calculates the error statistics using a Monte Carlo method. Both techniques can approximate the results obtained from the Kalman filter but at a much lower cost. The first technique, the RRSQRT, has been used throughout this work, while the second one, the EnKF, has been used only in the comparison study between both methods. For this reason, only the specific features of the RRSQRT implementation in the numerical model have been presented, though for the EnKF nearly the same features could be used.

A very important issue in every data assimilation application is the definition of the system noise. Two main sources of error were considered: errors in the open boundaries and errors in the meteorological forcing, although in some cases, errors in the continuity equation (water levels) were also considered. In the case of error in the meteorological forcing, and because these variables are not directly included in the state vector, the propagation equation of the error covariance matrix has to be modified. In the case of white noise, an additional cost in model run equivalents equal to the number of noise points is necessary. For the case of time-colored noise (augmented state vector formulation) a simplification has been introduced that allows the error covariance propagation equation to be solved without any additional cost. Another additional problem associated with a RRSQRT implementation is related to the different scales of magnitude of the variables in the state vector. In this work the eigenvalue decomposition is calculated on a normalised error covariance matrix. The error covariance matrix is normalised for each variable independently. The last two problems associated with the RRSQRT filter are not present in the EnKF due to its special way of defining the errors by using an ensemble.

The implementations in a two-dimensional hydrodynamic model of the two aforementioned data assimilation schemes have been tested and their performance compared using a twin experiment. The model noise process has been related to model errors through the open boundary conditions and the meteorological forcing. The use of an augmented state vector formulation provides simultaneous corrections of the system state and the forcing terms during the assimilation procedure.

The first conclusion extracted from the tests is that the most important parameter is the rank reduction for the RRSQRT filter and the number of ensemble members for the EnKF. It is very difficult to define apriori a value of this variable that will produce good filter results. The reduced rank representation of the RRSQRT is more efficient than the ensemble representation in the EnKF, though in terms of computational load this advantage is lost specially due to the eigenvalue calculation in the RRSQRT filter. Both filters proved to be very efficient. Using a very small number of leading eigenvalues or ensemble members compared with the dimension of the state vector, the filters are able to provide a very good estimation of the error in the system. In general, both filters have shown a significant robustness with respect to the incorrect specification of the error statistics.

Under these conditions the filters can still provide good model corrections, but the estimation of the system error as well as the estimation of the errors in the forcings deteriorates significantly. The use of an augmented state vector defining the noise as time coloured noise provides a more efficient algorithm and consequently a better filter performance than using a white noise definition. Especially in the case of biased errors, the filter gives a good estimate of the biased error even if the error statistics have not been correctly specified. Moreover, the special case of errors in the initial conditions (meaning for example a smaller initial volume) has been investigated. Under these conditions the filter is unable to correct the model if only errors in the momentum equation are considered. In order to correct the model at the measurement positions, the results in the rest of the model deteriorate. This problem will not be present if the model has significant open boundaries because the initial volume can be modified through these.

The implementation of the RRSQRT filter in a two-dimensional hydrodynamic model has been tested in a North Sea model for a period that covers the special storm event of February 1993. Water levels have been assimilated at 10 stations, while another three were kept for validation. During hindcasting, the filter is able to produce very good corrections at the measurement locations and moreover it produces a significant improvement in the validation stations. However, the filter does not produce a good estimation of either the error covariance or the error in the forcing in this case. This is a consequence of the simplistic definition of the noise i.e. constant in space and correlated in time through a simple AR(1) model. This definition is very far from what is present in reality. Further efforts have to be made in order to define more accurately the noise processes, though it is a very difficult task because usually the only information available is provided by a few measurements. During hindcasting the filter estimates the errors in the forcings in the augmented part of the state vector. Because the error could not be correctly estimated, as has been explained above, the use of this information during forecasting causes a deterioration in the forecasting results. This is not only a consequence of a badly estimated error but also because the AR(1) process used to propagate this error cannot represent the real error process. The usual approach of data assimilation for initialisation of the model prior forecast has been used. Therefore, the forecast is purely deterministic. The forecast results show that important improvements are achieved in some parts of the model during the first 6 hours, and up to 12 hours in the western Danish coast. However, the region most influenced by the northern boundary (the northern Scottish coast) deteriorates as a consequence of a sudden change from the corrected boundary to the deterministic boundary.

An important conclusion at this point is that the application of very sophisticated data assimilation techniques with a simple definition of the error structure and error statistics has an important limitation in state corrections as well as forecasting capabilities. An additional conclusion from this test was presented when a non-linear definition of the open boundary was used. In the case when the flux along the boundary is calculated by extrapolation from internal points during outflow and set to zero during inflow, a discontinuity in this variable is present. The RRSQRT filter became unstable in the region close to the boundary. This result was expected since the RRSQRT filter is based on the Extended Kalman filter, which is very well suited for linear and weakly non-linear dynamics, but can fail for strong non-linearities.

Furthermore, the RRSQRT filter has been implemented in a model that computes the two-dimensional shallow water equations in areas with different spatial resolution. The number of computational grid points grows very rapidly using a finer grid. The state vector has been defined in the main grid (the coarsest one). The aim of this approach is to obtain a filter that corrects in areas of finer resolution while maintaining similar computational time and storage requirements to those of a coarser grid. The Kalman gain, which is calculated in the main grid, is interpolated into the finer areas. A model of the North Sea and the Baltic Sea has been used for testing. A finer grid resolution has been included in the model in the inner Danish waters. The filter performance between the two models using only one area and using two areas have been compared. In both cases the filter produces a significant improvement in the 14 measurement locations and also an improvement of about 50% in the validation stations. Despite the finer resolution, the results did not improve in this area but deteriorated slightly. This is a consequence of two factors: the finer grid was not correctly calibrated in some areas and, most importantly, the direct interpolation of the errors from the coarse to the fine grid can generate important errors when the bathymetry is very irregular. The method developed for the nested two-dimensional model is more expensive than the one developed for a single area model only because the propagated model is more expensive, i.e. it contains more computational points. In this case when using the same number of leading eigenvalues with one or two areas, very similar results where obtained. The method of using a reduced state vector in the finer areas will probably fail if a third level resolution area is considered.

In the last application, the RRSQRT filter has been implemented into a three-dimensional model, which solves the governing equations under the assumption of hydrostatic pressure. The state vector in this initial implementation only contains the water levels and the two components of the depth averaged velocity, though the error covariance matrix is propagated using the full model equations. Under the conditions of this implementation, only the barotropic part is corrected and therefore the baroclinic effects cannot be corrected. The filter corrects the water levels and the depth averaged velocities and therefore indirectly the velocity profile and the vertical velocities. The implementation has been tested in a twin test for correcting biased model errors. In the case of biased model errors in the boundary conditions, the filter is able to correct very accurately all the model variables. This is due to the fact that the vertical profile depends on the water level variation, which is quite correctly estimated since water level data are assimilated. In the second case, a biased error in the meteorological forcing (either wind field and/or pressure field) is considered. Here the water levels are quite accurately estimated, however, despite these corrections, the corrections in the velocity profiles and vertical velocities are not as good as in the first case. In this case the velocity profiles depends strongly on the wind speed, which has to be fully estimated by the filter.

In order to correct the baroclinic effects and also if for example salinity data has to be assimilated, these variables have to be incorporated into the state vector. The same holds for observations of velocity profiles. They contain a lot of information but it can only be assimilated if these variables are included in the state vector. The assimilation of of salinity, temperature and velocity profile measurements needs to be attempted. For this purpose, either the inclusion of all the model variables in the state vector or some other suitable way of assimilating the vertical profiles will be considered. Future research will be conducted in this direction.

The sub-optimal schemes of the Kalman filter and specially the RRSQRT filter have proven to be very efficient methods for the assimilation of data into hydrodynamic numerical models. The cost associated with the RRSQRT application is very small when compared with the fully extended Kalman filter. One drawback of the RRSQRT is that the noise has to be related to variables included in the state vector, making it difficult to define the noise related to some external forcing, such as the meteorological forcing. Moreover the eigenvalue computation and the multiplication of the transpose of the square root approximation of the error covariance matrix by itself could be the most computationally expensive parts in the algorithm. If the number of variables in the state vector is very large, this multiplication becomes very costly. The most efficient methods for computing these two operations should be used. Finally, the multivariate nature of the state vector is a problem for the eigenvalue decomposition. A normalisation is required prior to the calculation of the eigenvalues. This problem has to be considered in the three dimensional case if the state vector contains variables such as velocities, salinity and temperature.

List of Figures

List of Tables

Bibliography

[1] M.B. Abbott, A. McCowan, and I.R. Warren. *Numerical modelling of free surface flows that are two-dimensional in plan.* Transport models for inland and coastal waters. Academic Press, London, 1981.

[2] M.B. Abbott and A.W. Minns. *Computational Hydraulics. Second Edition.* Avebury, Aldershot UK and Bookfield, USA, 1997.

[3] R.A. Anthes. Data assimilation and initialization of hurricane prediction models. *Journal of Atmospheric Sciences*, 31:702–719, 1974.

[4] L Bode and T.A. Hardy. Progress and recent developments in storm surge modeling. *Journal of Hydraulic Engineering ASCE*, 123–No.4:315–331, 1997.

[5] K. Bolding. Using a Kalman filter in operational storm surge prediction. In *Second International Symposium on Assimilation of Observations in Meteorology and Oceanography*, pages 379–383. World Meteorological Organization, March 1995.

[6] H. Buchard. 3D shallow water equations with a generalized vertical coordinate. Technical report, International Research Centre for Computational Hydrodynamics, ICCH, 1996.

[7] G. Burgers, P.J. Van Leeuwen, and G. Evensen. On the analysis scheme in the ensemble Kalman filter. *Monthly Weather Review*, 1996. Submitted Dec 1996.

[8] R. Cañizares. Data assimilation and parameter estimation in a 2-D advection-dispersion model. Master's thesis, International Institute for Infrastructural, Hydraulic and Environmental Engineering, Delft, The Netherlands, 1995. H.H. 222.

[9] R. Cañizares, A.W. Heemink, and H.J. Vested. Sequential data assimilation in fully non-linear hydrodynamic model. In *Hydroinformatics '96*. Balkema, The Netherlands.

[10] R. Cañizares, A.W. Heemink, and H.J. Vested. Application of advanced data assimilation methods for the initialisation of storm surge models. *Journal of Hydraulic Research*, 36–No.4:655–674, 1998.

[11] C.K. Chui and Chen. *Kalman filter with real-time applications*, volume 17 of *Spring Series in Information Sciences*. Spring-Verlag, 1991.

[12] S.E. Cohn and D.F. Parrish. The behavior of forecast error covariances for a Kalman filter in two dimensions. *Monthly Weather Review*, 119:1757–1785, 1991.

[13] S.E. Cohn and R. Todling. Approximate data assimilation schemes for stable and unstable dynamics. *Journal of the Meteorological Society of Japan*, 74:63–75, 1996.

[14] R. Daley. *Atmospheric data analysis*. Cambridge university press, Cambridge, U.K., 1991.

[15] P. De Mey. Optimal interpolation in a model of the Azores Current in 1986-88. In P.Brasseur and C.J. Nihoul, editors, *Data assimilation: tools for modelling the ocean in a global change perspective*, volume 47 of *NATO ASI*. Springer-Verlag, Berlin, 1994.

[16] P. De Mey. Data assimilation at the oceanic mesoscale: A review. *Journal of the Meteorological Society of Japan*, 71–1B:415–427, 1997.

[17] D.P. Dee. Simplification of Kalman filter for meteorological data assimilation. *Q.J.R. Meteorological Society*, 117:365–384, 1991.

[18] J.P. Delhomme. Kriging in hydrosciences. *Advanced in Water Resources*, 1:251–266, 1978.

[19] DHI. Mike 21 user guide and reference manual. Danish Hydraulic Institute., 1995. Denmark.

[20] M. Eknes and G. Evensen. Parameter estimation solving a weak constraint variational formulation for an ekman model. *Journal of Geophysical Research*, 102–C6:12479–12492, 1997.

[21] G. Evensen. Using the extended Kalman filter with a multilayer quasi-geostropic ocean model. *Journal of Geophysical Research*, 97–C11:17905–17924, 1992.

[22] G. Evensen. Sequential data assimilation with a nonlinear quasi-geostrophic model using Monte Carlo methods to forecast the error statistics. *Journal of Geophysical Research*, 99–C5:10143–10162, 1994.

[23] G. Evensen. Advanced sequential methods with nonlinear dynamics. In *Second International Symposium on Assimilation of Observations in Meteorology and Oceanography*, pages 147–158. World Meteorological Organization, March 1995.

[24] G. Evensen. The ensemble Kalman filter. Paper prepared for the Advanced School on Ocean Forecasting, IMC-Centro Marino Internazionale, July 1997. Torregrande, Sardinia.

[25] G. Evensen and P.J. Van Leeuwen. Assimilation of geosat altimeter data for the Agulhas current using the ensemble Kalman filter with a quasi-geostrophic model. *Monthly Weather Review*, 124:85–96, 1996.

[26] I. Fukimori and P. Malanotte-Rizzoli. An approximate Kalman filter for ocean data assimilation: an example with an idealized Gulf Stream model. *Journal of Geophysical Research*, 100–C4:6777–6793, 1995.

[27] A. Gelb. *Applied Optimal Estimation*. The M.I.T. Press, Cambridge, Massachusetts, and London, England, 1974.

[28] M. Ghil and P. Malanotte-Rizzoli. Data assimilation in meteorology and oceanography. *Advanced in Geophysics*, 33:141–266, 1991.

[29] K. Haines. A direct method of assimilating sea surface height data into ocean models with adjustments to the deep circulation. *Journal of Physical Oceanography*, 21:843–868, 1991.

[30] K. Haines, P. Malanotte-Rizzoli, W.R. Holland, and R.E. Young. A comparison of two methods for the assimilation of altimeter data into a shallow water model. *Dynamics of the Atmosphera and the Ocean*, 17:89–133, 1993.

[31] A.W. Heemink. *Storm surge prediction using Kalman filtering*. PhD thesis, Twente University of Technology, The Netherlands, 1986.

[32] A.W. Heemink. Two dimensional shallow water flow identification. *Applied Mathematical Modelling*, 12:109–118, 1988.

[33] A.W. Heemink. Identification of wind stress on shallow water surfaces by optimal smoothing. *Stochastic Hydrology and Hydraulics*, 4:105–119, 1990.

[34] A.W. Heemink, K. Bolding, and M. Verlaan. Storm surge forecasting using Kalman filtering. *Journal of the Meteorological Society of Japan*, 75–No1B:195–208, 1997.

[35] A. H. Jazwinski. *Stochastic Processes and Filtering Theory*. Academic Press, New York, 1970.

[36] H.R. Jensen. Dynocs technical report, regional model. Technical report, European Community research project MAS2-CT94-0088, 1997.

[37] R.E. Kalman. A new aproach to linear filtering and prediction theory. *J. Basic. Engr.*, 82D:35–45, 1960.

[38] R.E. Kalman and R.S. Bucy. New results in linear filtering and prediction theory. *J. Basic. Engr.*, 83D:95–108, 1961.

[39] R.W. Lardner, A.H. Al-Rabeh, and N. Gunay. Optimal estimation of parameters for a two-dimensional hydrodynamical model of the Arabian Gulf. *Journal of Geophysical Research*, 98–C10:18229–18242, 1993.

[40] J.J. Leendertse. *Aspects of a Computational Model for Water Wave Propagation*. M.I.T. Instrumentation Laboratory, Memo SGA 5-64, Cambridge, Massachusetts, 1964.

[41] B.P. Leonard. The ULTIMATE conservative difference shcheme applied to unsteady one-dimensional advection. *Comput. Methods Appl. Mech.*, 88:17–74, 1991.

[42] P.F.J. Lermusiaux. *Error Subspace data assimilation methods for ocean field estimation: theory, validation and application*. PhD thesis, Harvard University, Cambridge, Massachussetts, USA, May 1997.

[43] A.C. Lorenc. A global three-dimensional multivariate interpolation scheme. *Monthly Weather Review*, 109:701–721, 1981.

[44] H. Madsen. Internal meetings of the International research Centre for Computational Hydrodynamics, Nov 1997. Denmark.

[45] H. Madsen. On the use of Monte Carlo simulation methods for data assimilation in Mike 21. Technical report, International Research Centre for Computational Hydrodynamics, 1997.

[46] G. de Marsily. *Quantitative Hydrogeology. Groundwater hydrology for engineers*. Academic Press, New York, 1986.

[47] P.S. Maybeck. *Stochastic Models, Estimation and Control*, volume 141-1 of *Mathematics in Science and Engineering*. Academic Press, New York, 1979.

[48] J.C. McWilliams. Modeling the oceanic general circulation. *Annu.Rev. Fluid Mechanics*, 28:215–248, 1996.

[49] J.D. Pietrazk and K. Bolding. *Towards a Coastal Ocean Prediction system of the Danish domestic waters*. Coastal Ocean Prediction. 1998.

[50] J.D. Pietrzak. The use of TVD limiters for forward-in-time upstream-biased advection schemes in ocean modeling. *Monthly Weather Review*, 126:812–830, 1998.

[51] J.D. Pietrzak, J.B. Jacobsen, H.J. Vested, H. Buchard, and O. Petersen. A three-dimensional hydrostatic model for coastal and shelf sea modelling. Technical report, International Research Centre for Computational Hydrodynamics, ICCH, 1998.

[52] N. Pinardi. Ocean numerical modelling: A historical point of view. Paper prepared for the Advanced School on Ocean Forecasting, IMC-Centro Marino Internazionale, July 1997. Torregrande, Sardinia.

[53] J.E. Potter. *W Matrix Augmentation*. Rand Memorandum, RH-5299-PR, Santa Monica, California, 1967.

[54] R.W. Preisendorfer. *Principal components analysis in meteorology and Oceanography.* Elsevier, Amsterdam, 1988.

[55] M.M. Rienecker and R.N. Miller. Ocean data assimilation using optimal interpolation with a quasigeostrophic model. *Journal of Geophysical Research,* 96:15093–15103, 1991.

[56] A. Sokolov, O. Andrejev, F. Wulff, and M. Rodriguez Medina. The data assimilation system for data analysis in the Baltic Sea. In *Systems Ecology contributions,* volume 3. Stockholm University, Stockholm, 1997.

[57] O. Talagrand and P. Courtier. Variational assimilation of meteorological observations with the adjoint vorticity equations. part i, theory. *Q.J.R. Meteorological Society,* 1988.

[58] W. Thacker and R.B. Long. Fitting dynamics to data. *Journal of Geophysical Research,* 93–C2:1227–1240, 1988.

[59] R. Todling and S.E. Cohn. Suboptimal schemes for atmospheric data assimilation based on the Kalman filter. *Monthly Weather Review,* 122:2530–2557, 1994.

[60] M. Verlaan. Some extensions to the calibration program WAQAD. Technical Report 94-111, Faculty of Technical Mathematics and Informatics, Delft University of Technology, 1994.

[61] M. Verlaan. Convergence of the RRSQRT algorithm for large scale Kalman filter problems. Technical Report 97-19, Faculty of Technical Mathematics and Informatics, Delft University of Technology, 1997.

[62] M. Verlaan. *Efficient Kalman filtering algorithms for hydrodynamic models.* PhD thesis, Technical University of Delft, The Netherlands, 1998.

[63] M. Verlaan and A.W. Heemink. Reduced rank square root filters for large scale data assimilation problems. In *Second International Symposium on Assimilation of Observations in Meteorology and Oceanography,* pages 247–252. World Meteorological Organization, March 1995.

[64] M. Verlaan and A.W. Heemink. Tidal flow forecasting using reduced rank square root filters. *Stochastic Hydrology and Hydraulics,* 11:349–368, 1997.

[65] H.J. Vested, W.J. Nielsen, H.R. Jensen, and K. Bolding. *Skill assessment of an operational hydrodynamic forecast system for the North Sea and Danish Belts,* volume 47 of *Coastal and Estuarine Studies.* American Geophysical Union, Washington, DC, 1995.

[66] J.W. De Vries. Verification of the WAQUA/CSM-16 model for the winters 1992/1993 and 1993/1994. Technical Report TR 176, KNMI, 1995.

[67] E.F. Wood and P.E. O'Connell. Real-time forecasting. In *Hydrological Forecasting*. Wiley, New York.

[68] X.F. Zhang. *Data assimilation in air pollution modelling*. PhD thesis, Delft University of Technology, Delft, The Netherlands, 1997.

[69] Y. Zhou. Kalmod a stochastic-deterministic model for simulating groundwater flow with Kalman filtering. Technical Report IHE-22, International Institute for Infrastructure, Hydraulic and environmental Engineering, Delft, The Netherlands, 1991.

Samenvatting
(Summary in Dutch)

Over data assimilatie toepassingen in regionale kustmodellen

In deze dissertatie wordt de ontwikkeling en integratie van bestaande rekenkundig efficiënte data assimilatie technieken voor grootschalige regionale hydrodynamische modellen beschreven. Het gebruik van verschillende numerieke modellen in de loop van dit onderzoek maakte het noodzakelijk om specifieke data assimilatie technieken te ontwikkelen voor elk van deze modellen.

In de inleiding wordt eerst een overzicht gegeven van de verschillende data assimilatie technieken en worden vervolgens de gebruikte numerieke modellen gepresenteerd. De vergelijkingen en de numerieke methoden toegepast in het twee-dimensionale model MIKE 21 en het drie-dimensionale model MIKE 3 HS worden hierbij ook beschreven.

Vervolgens worden het lineaire en het extended Kalman filter geïntroduceerd. Ook wordt een definitie van het Kalman filter met gebruikmaking van een uitgebreide toestandsvector formulering gegeven. Twee suboptimale oplossingsmethoden worden beschreven. Het eerste oplossingsschema is een Reduced Rank Square Root (RRSQRT) benadering van het extended Kalman filter, dat de covariantie matrix van de schattingsfout benadert met een matrix van een lagere rangorde. Het tweede oplossingsschema is het Ensemble Kalman Filter (EnKF) dat de foutstatistiek berekent met een Monte Carlo methode.

De hierboven genoemde data assimilatie schema's zijn getest in een twee-dimensionaal hydrodynamisch model en hun gedrag is vergeleken door middel van een tweeling experiment. Het modelruis proces wordt hierbij in verband gebracht met modelfouten voortkomend uit de open randvoorwaarden en de meteorologische forcing. Het gebruik van een uitgebreide formulering van de toestandsvector in de assimilatieprocedure levert gelijktijdige correctie van zowel de systeemtoestand als van de forcing termen op.

Het RRSQRT filter in een twee-dimensionaal hydrodynamisch model is getest bij een Noordzee model gedurende een periode in 1993, die o.a. de storm van februari 1993 omvat.

Het RRSQRT filter is ook geïmplementeerd in een model dat de twee-dimensionale ondiepwatervergelijkingen simuleert met verschillende roosters in afzonderlijke deelgebieden. Het aantal berekeningspunten in een model neemt zeer snel toe met het verfijnen van het rooster. Het doel van deze benadering is om ook te kunnen corrigeren in de gebieden met het fijne rooster, maar tegelijkertijd de rekentijd en het geheugengebruik te beperken.

In de laatste toepassing is het RRSQRT filter geïmplementeerd in een drie-dimensionaal model dat de primitieve vergelijkingen simuleert onder de aanname van hydrostatische druk. De toestandsvector omvat in deze toepassing alleen de waterstanden en de twee componenten van de over de diepte gemiddelde stroomsnelheid, hoewel de covariantie matrix van de schattingsfout op de volledige vergelijkingen toegepast wordt. In deze toepassing kan alleen voor de barotrope component gecorrigeerd worden en niet voor de barocline effecten. Omdat het filter zowel de waterstanden als de gemiddelde horizontale stroomsnelheden corrigeert, verbetert het indirect ook de snelheidsprofielen en de verticale snelheden.

Acknowledgements

I would like to express my gratitude to all the persons who have, in different ways, contributed to the success of this research.

This work has been financed by the Danish National Research Foundation through the International Research Centre for Computational Hydrodynamics. Their support is sincerely appreciated.

My next words are dedicated to my promotors Prof. Mike B. Abbott and Prof. Arnold W. Heemink. Prof. M.B. Abbott wisely guided me through the field of computational hydraulics and hydroinformatics. Prof. A.W. Heemink was a continuous source of inspiration for my understanding of data assimilation. I deeply appreciate their continuous support.

Very special thanks are due to Ir. H.J. Vested, head of the Ecological Modelling Centre of the Danish Hydraulic Institute. His interest in data assimilation, and his constant support and perseverance made this research possible.

Many important ideas and suggestions for a research were born during discussions with colleagues. I should like to thank Dr. Ir. M. Verlaan from the Department of Applied Mathematics and Informatics at the Delft University of Technology and Dr. H. Madsen from the Danish Hydraulic Institute for our uncountable discussions that were an endless source of ideas.

I would like to express my gratitude to all my colleagues from the Danish Hydraulic Institute and the International Research Centre for Computational Hydrodynamics who have supported this work, with a special thanks to H.R. Jensen, J.B. Jacobsen and Dr. Julie D. Pietrazk. I should also like to thank my IHE Ph.D. college Juan C. Savioli, with whom I shared an office during most of my research period, for his company, support and friendship.

To my family and specially my parents, my heartfelt thanks and love. They always support my dreams and help me to carry them out. I am so grateful to my wife Eugenia that is difficult to express it with words. Without her continues love, help and support this work would not have been possible.

Curriculum Vitae

Rafael Cañizares was born on March 28, 1967 in Madrid, Spain. In 1993 he received the degree of M.E. Civil Engineer (Ingeniero de Caminos, Canales y Puertos) from the Polytechnical University of Madrid. As part of the Ph.D. programme, he attended in 1993-94 the postgraduate course on Hydroinformatics at the International Institute for Infrastructural, Hydraulic and Environmental Engineering (IHE) in Delft, The Netherlands and obtained the Diploma of Hydraulic Engineering (Dip. HE Delft) with distinction. In April 1995 Rafael Cañizares obtained the Master of Science degree with distinction at IHE Delft on the basis of his work carried out at the International Research Centre for Computational Hydrodynamics and Danish Hydraulic Institute (ICCH / DHI) with a dissertation entitled *Data Assimilation and Parameter Estimation in a 2-D Advection-Dispersion Model*. Until December 1997 he worked as a Research Fellow towards his Ph.D. degree at the International Research Centre for Computational Hydrodynamics and Danish Hydraulic Institute (ICCH / DHI), where most of the research included in this thesis was carried out.

The aim of the International Institute for Infrastructural, Hydraulic and Environmental Engineering,
IHE Delft, is the development and transfer of
scientific knowledge and technological know-how
in the fields of transport, water and the environment.

Therefore, IHE organizes regular 12 and 18 month
postgraduate courses which lead to a Masters Degree.
IHE also has a PhD-programme based on research,
which can be executed partly in the home country.
Moreover, IHE organizes short tailor-made and
regular non-degree courses in The Netherlands as
well as abroad, and takes part in projects in various
countries to develop local educational training and
research facilities.

International Institute for
Infrastructural, Hydraulic and
Environmental Engineering

P.O. Box 3015 –
2601 DA Delft
The Netherlands

Tel.:   +31 15 2151715
Fax: +31 15 2122921
E-mail: ihe@ihe.nl
Internet: http://www.ihe.nl